Respiratory Physiology

A Clinical Approach

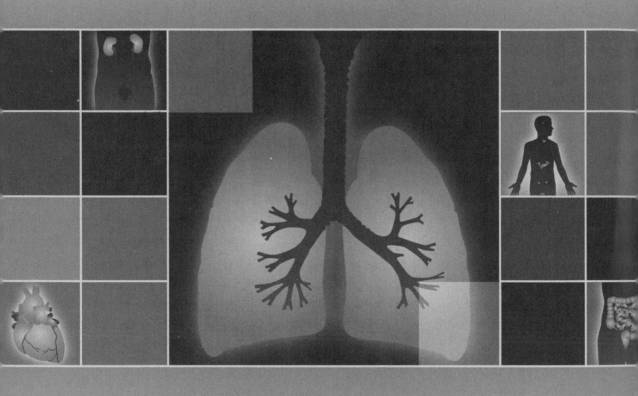

Respiratory Physiology
A Clinical Approach

Richard M. Schwartzstein, MD
Clinical Director, Division of Pulmonary and Critical Care Medicine
Vice President for Education
Beth Israel Deaconess Medical Center
Director, Carl J. Shapiro Institute for Education
Associate Professor of Medicine
Faculty Associate Dean for Medical Education
Harvard Medical School
Boston, MA

Michael J. Parker, MD
Instructor in Medicine
Division of Pulmonary and Critical Care Medicine
Beth Israel Deaconess Medical Center
Senior Interactive Media Architect
Center for Educational Technology
Harvard Medical School
Boston, MA

LIPPINCOTT WILLIAMS & WILKINS
A **Wolters Kluwer** Company
Philadelphia • Baltimore • New York • London
Buenos Aires • Hong Kong • Sydney • Tokyo

Editor: Betty Sun
Managing Editor: Cheryl W. Stringfellow
Developmental Editor: Kathleen H. Scogna
Marketing Manager: Emilie Linkins
Production Editor: Sirkka E. H. Bertling
Designer: Doug Smock
Contributing Educator: Nurit Bloom, Harvard Medical School
Compositor: Nesbitt Graphics
Printer: Courier–Kendallville

Library of Congress Cataloging-in-Publication Data

Schwartzstein, Richard M.
    Respiratory physiology : a clinical approach / Richard M. Schwartzstein, Michael J. Parker.
        p. ; cm.
    Includes index.
    ISBN 0-7817-5748-7
    1. Respiration. 2. Respiratory organs--Physiology. I. Parker, Michael J. II. Title.
    [DNLM: 1. Respiratory Physiology. WF 102 S399r 2005]
QP121.S285 2005
612.2--dc22   2005007585

To purchase additional copies of this book, call our customer service department at **(800) 638-3030** or fax orders to **(301) 824-7390**. International customers should call **(301) 714-2324**.

Visit Lippincott Williams & Wilkins on the Internet: http://www.LWW.com. Lippincott Williams & Wilkins customer service representatives are available from 8:30 am to 6:00 pm, EST.

05 06 07 08 09
1 2 3 4 5 6 7 8 9 10

To my grandparents, Eva and Getzl, and my parents, Muriel and Sam: you never stopped believing in the opportunities for your family and in the potential we all possess.

—RMS

To my parents, Leonard and Gloria: the embodiment of warmth, curiosity, scientific passion and integrity, and above all, steadfast support.

—MJP

Preface

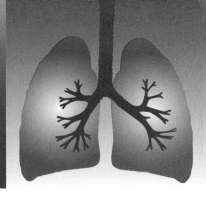

The goal of *Respiratory Physiology: A Clinical Approach* is to provide a clear, clinically oriented exposition of the essentials of respiratory physiology for medical students, residents, nurses, and allied health professionals. We present the physiology in the context of a system to emphasize that the functions we associate with breathing depend on more than the lungs. This approach is essential for a complete understanding of the clinical problems that affect respiration and that lead to symptoms of shortness of breath.

This book is the first in a series of monographs on physiology. The series will address cardiovascular, renal, gastrointestinal, and endocrine function, in addition to respiratory physiology. Each book will be designed to address the learners outlined above and will use the same style and pedagogical tools. Our goals are to present physiology in a clinically meaningful way, to emphasize that physiology is best understood within the context of an organ *system*, to demonstrate principles common to different systems, and to use an interactive style that engages and challenges readers.

Level

The level of the book is intended to fit a range of needs from students who have had no previous exposure to physiology to residents who are now in the thick of patient care but feel the need to review relevant physiology in a clinical context. We have drawn upon many years of experience teaching students, residents, and fellows in making decisions with respect to the topics emphasized and the clinical examples used to illustrate key concepts. The book is not intended as a comprehensive review of respiratory physiology nor is it designed for advanced, research oriented physiologists. Rather, we have focused on issues that are most relevant for the care of patients while, at the same time, providing sufficient physiological detail to provide readers with the foundation to examine and analyze new data on these topics in the future.

Most of the concepts presented in the book are well established, and we do not burden you with reference lists for this information. When we present newer and, in some cases, more controversial issues, however, we do provide relevant primary source citations.

Content

The book begins with two chapters that serve to provide context for the study of respiratory physiology. In Chapter 1, we lay out the framework (controller, ventilatory pump, and gas exchanger) of the respiratory system upon which we will hang concepts and information in the succeeding chapters. In the process of building this framework, we also give you an opportunity to experience some of the interactive elements of the text that will be revisited in subsequent chapters. Chapter 2 focuses on functional anatomy, reviewing the essential elements of which the respiratory system is composed and linking the structure of these elements to their physiological role.

The next two chapters address what historically has been called "respiratory mechanics." Chapters 3 and 4 focus on the function of the ventilatory pump, first under conditions when there is no flow in or out of the lungs (statics) and then when the system is set in motion (dynamics). Chapter 5 centers on the gas exchanger and delineates essential principles necessary for an understanding of the factors that affect the movement of oxygen and carbon dioxide into and out of the blood.

Chapter 6 focuses on the controller, the neurological regulation of breathing. This is linked closely with the next two chapters, which provide an introduction to acid–base physiology and respiratory sensations. Although a complete understanding of acid–base physiology depends on knowledge of renal function, the respiratory system is a critical player in the body's response to alterations in pH, and diseases of the respiratory system are often the cause of acid–base abnormalities. The fundamentals of acid–base physiology are addressed in Chapter 7. Chapter 8 discusses the physiology of respiratory sensations. This topic is not traditionally covered in introductory physiology texts, but we have included it because it forms the basis for an appreciation of the symptoms of breathing discomfort found in a variety of disease states. Much of the information provided in this chapter is derived from research conducted in the past 2 to 3 decades and is less well disseminated among practicing physicians than is the content in the remainder of the book. For this reason, we have provided a reference list at the end of this chapter.

We conclude with Chapter 9, exercise physiology, which integrates the materials from the entire book and demonstrates links between respiratory and cardiovascular physiology. This chapter may be challenging for you if you have not yet studied cardiovascular physiology, but it will provide you with an introduction to that system upon which you can build in the future.

Throughout the book, we draw heavily on clinical examples to emphasize concepts and to highlight how an understanding of normal physiological principles will help you understand pathological states. Beginning students will see the relevance of the material presented. Advanced students and residents will be able to use the book's clinical examples to understand the signs and symptoms of your patients and the rationales for therapeutic interventions.

Pedagogy

- **Chapter outline:** The outlines at the beginning of the chapters provide previews of the chapters and are useful study aids.
- **Learning Objectives:** Each chapter starts with a short list of learning objectives, which are intended to help you focus on the most critical concepts and physiological principles that are presented in the chapter.
- **Text:** The text is written in a conversational style that is intended to recreate the sense of participating in an interactive lecture. Questions are posed periodically to offer you opportunities to reflect on information presented and to try your hand at synthesizing and applying your knowledge to novel situations.
- **Topic headings:** Topic headings are used to delineate key concepts. Sections are arranged to present the material in easily digestible quantities as you move from simple to more complex physiology.
- **Boldfacing:** Key terms are boldfaced upon their first appearance in a chapter. Definitions for all boldfaced terms are found in the glossary.

- **Thought Questions**: Interposed within the text are Thought Questions that are designed to challenge you to use the material just presented in the text in a novel fashion. Many of these questions are posed in a clinical context to demonstrate the clinical relevance of the material as well. We encourage you to try these questions on your own before checking the answers that appear at the end of each chapter.

- **Illustrations and Animated Figures**: The figures have been developed to demonstrate the relationships between physiological variables, to illustrate key concepts, and to integrate a number of principles enumerated in the text. Many of the figures in this book are linked to interactive learning tools (called Animated Figures in the text and indicated by an icon) that will provide you with an opportunity to view a physiological principle in motion or to manipulate variables and see the physiological consequences of the changes. These animations and computer simulations permit you to work with the concepts and apply them in a range of circumstances. Our goal is that, with self-paced use of these interactive animations, you will gain a deeper, more intuitive understanding of the physiological principles discussed in each chapter.

- **Tables and Quick Checks**: Tables and Quick Checks provide summaries of information outlined in the text.

- **"Putting It Together" sections**: At the end of each chapter is a clinical case presentation that poses questions about physical findings, laboratory values, or diagnostic and therapeutic issues that can be answered with the physiological information presented in the chapter. These cases are designed to integrate material, to demonstrate the clinical relevance of the physiology, and to provide you with an opportunity to test yourself by applying what you have just learned in a new situation.

- **Review Questions and answers**: You can use the Review Questions at the end of each chapter to test whether you have mastered the material. For medical students, the USMLE-type questions should help you prepare for the Step 1 examination. Answers to the questions are presented at the end of the book, and include explanations that delineate why the choices are correct or incorrect.

- **Index**: A complete index allows you to easily find material in the text.

In the final analysis, most people study physiology because it offers great insights into the workings of the human body. We have organized and presented the materials in this book in a way that we hope will allow you to achieve your individual goals while having some fun with a subject that continues to challenge and intrigue us.

Richard M. Schwartzstein, MD
Michael J. Parker, MD

Acknowledgments

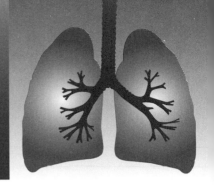

As with most academic endeavors, this one is the product of work that has been inspired and shaped by many people. We thank Steven Weinberger, MD, whom I (RS) am proud to call a mentor and friend, who gave me the opportunity to teach respiratory physiology at Harvard Medical School and encouraged this project from its inception until its conclusion. J. Woodrow Weiss, MD, who helped pull me (RS) from cardiology to pulmonary and critical care medicine during my internship, provided considerable emotional support throughout the development and writing of the book. Steven Loring, MD, one of the best physiologists and teachers we know, provided important suggestions on an early draft of Chapter 3 and was always available to answer questions to ensure that our efforts to simplify concepts to a level appropriate for students did not lead to any deviations from truth.

John Halamka, MD, had the vision to encourage the development of concept simulation in the Harvard curriculum and supported our efforts to build on those foundations in the use of technology in this project. Nurit Bloom provided extraordinary guidance and assistance in the development of tools for the testing of the animations and simulations with medical students and was a careful reader of the text for inconsistencies and for areas that needed greater clarity.

We thank Betty Sun of Lippincott Williams & Wilkins for her appreciation of the potential of this project from its beginnings and for her support during a tight writing and editing schedule. Liz Allison, our agent and confidante, was always there for us. She has become a dear friend during the many months we labored on this book, the completion of which is a testament to her persistence, encouraging words, and faith in us.

To our past students, we thank you for the superb questions you asked that made us think again about the principles underlying respiratory physiology and compelled us to develop new ways of conveying these complex concepts. To our future students, we ask that you assist us in making this text even better.

Finally, we would like to thank our families. To my (RS) children, Pete, Nick, Jake, and Hannah, thank you for sharing me with my students and patients from whom I have learned so much over the years. To my wife, Sarah, I owe an incalculable debt of gratitude for supporting me equally through the highs and lows that accompany such a project and for creating the warm and loving environment that has enabled me to get to this point. To my (MP) parents, Leonard and Gloria, thank you for unending interest and encouragement in this project and everything leading up to it. Your unconditional support and love have always meant more to me than you can know.

Contents

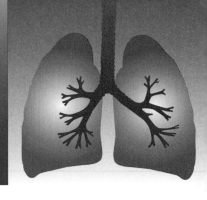

Getting Started: The Approach to Respiratory Physiology

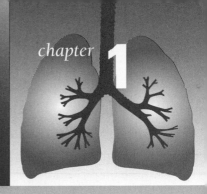

CHAPTER OUTLINE

LEARNING OBJECTIVES

- **To develop the concept that respiratory physiology must be studied within the context of the respiratory system.**
- **To outline the elements of the respiratory system.**
- **To delineate the learning tools that will be used throughout this book.**
- **To describe the best methods for use of the learning tools contained within this book.**

The act of breathing is essential to life, and the study of the physiology of the respiratory system is relevant for all medical specialties from internal medicine to surgery to psychiatry. An understanding of gas exchange, the movement of oxygen into the blood to support aerobic metabolism in the mitochondria and removal of the metabolic byproduct, carbon dioxide, from the blood, is critical for all physicians providing clinical care. The multiple factors that can interfere with this process cut across specialties. The role of the respiratory system in maintaining equilibrium of the acid–base balance of the body is another essential topic for all practitioners. The unique blend of automatic and volitional control of breathing makes the understanding of human behavior a relevant issue as well. If you plan on entering the field of psychiatry some day, for example, you may be confronted with a patient with panic attacks, hyperventilation, and a series of symptoms that can best be understood within the context of respiratory physiology.

When you first heard you were about to study respiratory physiology, chances are that your first thought was: "I will be studying the lungs." Although the lungs are clearly an essential element of the processes we associate with breathing, they alone do not provide the whole story. Rather, one must step back and look at the *system* that is responsible for the movement of oxygen into the blood and carbon dioxide out. Without muscles to generate a negative pressure within the thorax to draw air into the lungs, for example, the lungs would be worthless. Without an area of the brain to monitor breathing and to send appropriately timed neurological impulses to the muscles and the peripheral nerves to carry those impulses, the muscles could not do their job. As you proceed from one chapter to another, we will periodically orient you to where we are in the respiratory system and show you how the different components interact.

At its core, physiology is a conceptual science. The study of physiology is critical to understanding the way that the body functions and the many mechanisms available to restore homeostasis when disease attacks. Our goal throughout this book is to emphasize conceptual understanding of the material. We want you to develop an appreciation of physiological principles at a depth sufficient to allow you to apply the concepts to new situations, thereby enabling you to make sense of a unique patient, the ultimate challenge in medicine. Our emphasis is on *clinical physiology*. Given the great range of information that physicians in training must learn, we emphasize principles that are most essential for you to care for patients and to have a strong foundation upon which to build as you move on to more advanced levels of study. Although we use clinical examples throughout the text to demonstrate the relevance of the material and the ways that the concepts are applied to patients, this is a physiology book, not a pathophysiology one. Thus, the book does not describe in detail specific disease states, diagnostic methods, or treatment options except as they may enhance your understanding of physiology. Given the emphasis on this clinical approach, we will not spend considerable time on some physiological points found in classic texts that are neither clinically relevant nor necessary to have a solid foundation in the field. We acknowledge that this is not a text aimed primarily at basic science physiologists. Rather, we strive to provide clinical physiologists with essential information needed for patient care today and the understanding to add to that knowledge in the future.

Much of what we know about physiology derives from careful observations of people and from animal models. From these observations, hypotheses have been constructed to help explain the findings. In some cases, hypotheses have been tested extensively and modified accordingly. In other situations, we are still working largely at the level of conjecture. Those who have a need for definitive answers in all circumstances may be frustrated at times in the study of physiology. The "proof" is not always available to us. Perhaps you may be intrigued by one of these areas of uncertainty and pursue further investigations to provide us all with greater insights into the workings of the human body.

This book is designed primarily for medical students who are embarking on the study of respiratory physiology for the first time, but the book also serves as an excellent review for advanced medical students, interns, and residents, especially for those whose initial instruction in respiratory physiology took a more traditional basic science approach to the subject. Nurses and respiratory therapists who care for patients with respiratory problems may also find this text useful. This book is not intended as a definitive or all-encompassing resource on respiratory physiology. Those pursuing advanced training in pulmonary and critical care medicine or anesthesiology may rely on this as a primer as they expand their knowledge into the subtleties of the field.

A Systems Approach to Respiratory Physiology

In teaching physiology to medical students for the past decade, we have often heard a plea for assistance in making the different things they were learning "fit together." For knowledge to be both meaningful and useful, it is important to have a superstructure upon which to hang individual concepts. Similarly, in teaching interns and residents for the past 20 years, we have seen many mistakes made in the care of patients with respiratory disease because of the exclusive focus on the lungs as the physician tried to analyze the problem at hand. The answer to both of these problems is a systems approach to respiratory physiology.

The respiratory system is composed of all of the elements needed to move air from the atmosphere down to the alveoli, exchange oxygen and carbon dioxide in the blood flowing through the pulmonary capillaries, and then move carbon dioxide back to the atmosphere. The core components of this system are the respiratory controller, ventilatory pump, and gas exchanger.

QUICK CHECK 1-1   **COMPONENTS OF THE RESPIRATORY SYSTEM**

Respiratory controller
Ventilatory pump
Gas exchanger

The **respiratory controller** consists of the elements of the central nervous system (CNS) that tell us how often and how deeply to breathe. The **ventilatory pump** is composed of multiple structures that perform the bellows function of the respiratory system.

QUICK CHECK 1-2   **THE VENTILATORY PUMP**

Chest wall muscles
Chest wall skeleton (ribs, cartilage, sternum, spine)
Chest wall connective tissue
Airways
Pleura
Spinal cord and peripheral nerves

The muscles, bones, cartilage, and related soft tissue of the chest wall are necessary to move the chest wall and create changes in intrathoracic pressure. The peripheral nerves that connect the CNS to the muscles are critical to transform the messages from the controller into movement of the chest wall. Airways serve as conduits for flow of gas from the mouth to the alveoli, and the pleura connect the motion of the chest wall to the lungs. Finally, the **gas exchanger**, the alveoli and the pulmonary capillaries, is the site of the exchange of oxygen and carbon dioxide (Fig. 1-1).

Chapter 2 provides an overview of the anatomy of the respiratory system, emphasizing the function of the anatomic structures. Next the book explores the workings of the ventilatory pump, both under static conditions (no flow of gas; Chapter 3) and dynamic conditions (when the system is set in motion; Chapter 4). Chapter 5 discusses the gas exchanger, and Chapter 6 examines the unique aspects of respiratory control. Acid–base physiology, which plays an important role in ventilatory control, is introduced in Chapter 7. The origins of dyspnea, or shortness of breath, one of the most debilitating symptoms in clinical medicine, are far more complex than was thought even just a few years ago. To help you understand dyspnea when you move on in your studies of pathophysiology, we will explore the physiologic bases of respiratory sensations, which draw on elements of all three components of the respiratory system, in Chapter 8. Finally, Chapter 9 integrates all that you have learned and makes some links from your knowledge of the respiratory system to the cardiovascular system, as we examine the physiology of exercise. As we develop the concepts in each of these chapters, we will continually bring you back to the

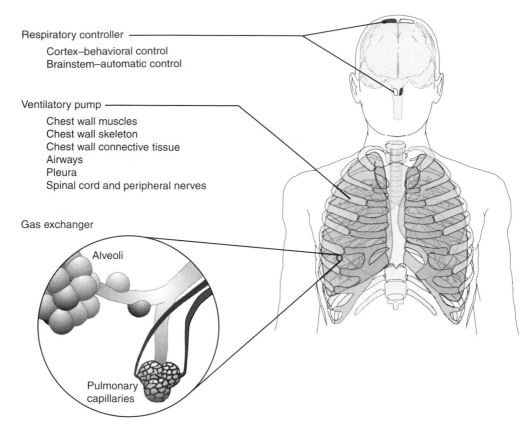

Respiratory controller
  Cortex–behavioral control
  Brainstem–automatic control

Ventilatory pump
  Chest wall muscles
  Chest wall skeleton
  Chest wall connective tissue
  Airways
  Pleura
  Spinal cord and peripheral nerves

Gas exchanger
  Alveoli
  Pulmonary
  capillaries

FIGURE 1-1 The components of the respiratory system. The respiratory system is composed of three key components: the controller, the ventilatory pump, and the gas exchanger. Each of these components has several elements. To achieve the functional goals of the respiratory system, each of these elements must be working effectively and in an integrated manner. To understand the way that diseases affect breathing and the ways that the body tries to compensate for these impairments and restore homeostasis, you must have a solid appreciation of the physiology of each of these parts.

superstructure of the respiratory system—controller, pump, gas exchanger—so you will understand how the individual pieces fit together.

A Clinical Approach to Respiratory Physiology

Given the vast quantities of information confronting physicians today, most students and residents want to know why a new piece of knowledge is essential for them to master. Why should I learn this? As a general principle, the answer to this question has two components. First, this knowledge will enable you to take better care of your patients today and in the future. Second, the construction of this foundation or framework will help you incorporate new knowledge about human function and disease as it is discovered in the future.

To help you see the relevance of respiratory physiology, we place the study of this subject within a clinical context as much as possible. We draw on clinical examples to reinforce and clarify concepts. At the end of each chapter, we provide a section called "Putting It Together" which presents a brief clinical scenario that illustrates many of the principles of the chapter and further reinforces the concepts just developed. Thus, the question of relevance is addressed in a very explicit manner.

The Keys to the Vault: Helping You Master the Material

As teachers, we want our students to be able to think critically; as students, you hope to be able to demonstrate critical thinking. To think critically about physiology and your patients, you must be able to understand concepts, not merely so they can be repeated but also in a way that they can be applied to new situations. To do this, you must achieve a level of understanding that is more than skin deep; you must have an intuitive feel for the concepts and their implications for the body's function. You must be able to manipulate the principles so they can shed insight on puzzles and open doors for the way out of the maze. To help you achieve this goal, we have taken an approach that is more conceptual than quantitative. We emphasize the equations and calculations you will need as a clinician and otherwise use numbers when doing so will enhance understanding rather than distract from it. We have provided a number of learning tools that are intended to help you develop the depth of understanding necessary for you to be able to think critically as a clinical physiologist.

ANIMATED FIGURES

To give you a chance to work with the concepts developed in the text, you will have opportunities to try a variety of computer-based animations and simulations. You can find the Animated Figures on the website or CD-ROM associated with this book. These interactive diagrams are designed to allow you to view an animation of a concept, or to manipulate, at your own pace, a number of variables that change over a range of physiologic conditions. You can experiment by changing parameters, predict the consequences of these changes, and then puzzle over which principles accounted for the transition from the first condition to the second. For visual learners (and for those who believe that "a picture is worth a thousand words"), these learning opportunities will enhance the text. In certain cases, we provide simultaneous auditory, textual, and animated material to engage your senses in the learning experience.

In some of the more complex Animated Figures, you should focus on one aspect of the animation at a time and then sit back and see if you can integrate everything that is happening at once. In essence, focus on one part of the puzzle and then another until the entire picture reveals itself to you. For example, take a look at Figure 1-2 and its associated

animated figure.

There is much going on here. Air is moving into and out of the lungs, and we see a visual manifestation of the movement of air on the volume–time trace with a device called a water spirometer. At the same time, we are looking at the forces being exerted by different elements of the chest wall and the lungs. How do these forces change as we go from a resting position at the end of exhalation to a maximal inspiration and then maximal expiration? You will need to ask yourself questions like this one when using the diagrams with the text. In certain cases, the text will pose these questions to you, guiding your use of the figure. Such a diagram may seem a bit overwhelming to you right now, but as you progress through the chapters, the concepts depicted will become clear to you.

THOUGHT QUESTIONS

In the early 20th century, the world-renowned physicist Albert Einstein created *gedanken experiments*, or thought experiments, to help him analyze the forces of the universe. These were experimental situations that he worked through intellectually rather than physically to find fallacies in his reasoning. In the spirit of Einstein, we have interspersed thought questions throughout the text to assist you in working through concepts and principles of respiratory physiology. We urge you to take the time to ponder the thought questions

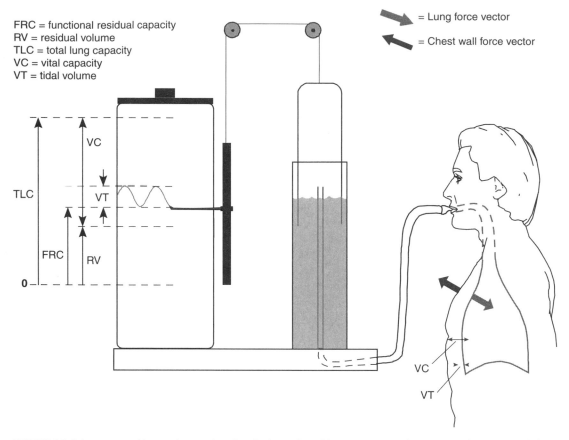

FRC = functional residual capacity
RV = residual volume
TLC = total lung capacity
VC = vital capacity
VT = tidal volume

= Lung force vector

= Chest wall force vector

VC

TLC   VT

FRC   RV

0

VC

VT

FIGURE 1-2 Spirometry and lung volumes. A patient is shown breathing on a water spirometer. As the patient exhales, the expired gas forces the spirometer drum up; during inhalation, the drum descends. The movement of the drum is transformed into movement of a pen, and the system is calibrated so that the vertical motion can be translated into changes in volume. The figure depicts the patient at a lung volume called *functional residual capacity*. Note that at this lung volume, inward force exerted by the lungs (the elastic recoil of the lungs) is equal in magnitude but opposite in direction to the outward recoil force exerted by the chest wall. **Also see Animated Figure 1-2.** This Animated Figure reappears with a fuller explanation in Chapter 3.

when they arise because they are strategically placed to reinforce the concepts developed up to that point in the text and Animated Figures. Thus, an inability to answer the thought question should prompt you to revisit the material that precedes it.

 THOUGHT QUESTION 1-1: Imagine that you are traveling through the galaxy at velocities approximating the speed of light. Which properties of the ventilatory pump would be more affected, the static (when viewing the respiratory system when you freeze it for a moment under conditions in which there is no flow) or dynamic (when viewing the respiratory system over time as air flows into and out of the lungs) properties? Why?

Answers to the thought questions can be found at the end of each chapter.

REVIEW QUESTIONS

At the end of each chapter, the review questions will enable you do a self-assessment of your learning. The questions are based on mini case scenarios. Answers to the review questions are found in an appendix at the end of the book. In addition, a glossary of terms is included at the end of the text to facilitate your learning of the vocabulary of respiratory physiology.

PUTTING IT TOGETHER

As you proceed through the chapters, you will learn a number of concepts and facts about different elements of the respiratory system. To allow you to see how the things you have learned can be used to analyze a patient problem, each chapter ends with a section called "Putting It Together." These sections present clinical scenarios and pose diagnostic or management questions. We then show you how your growing knowledge of physiology can be applied to the case to understand the particular patient's problem. As previously noted, this is not a pathophysiology book, and we do not expect you to have knowledge of the disease processes themselves. We do believe, however, that it is helpful to see how physiological principles are used on a daily basis in patient care. Even though you have just scratched the surface of respiratory physiology in this introductory chapter, let's try a case together to show you how this works:

A 33-year-old man is found on the street and is brought to the emergency department. He responds only to painful stimuli, smells strongly of alcohol, and has shallow, slow breathing. The physical examination reveals evidence of several fractured ribs in the right chest. The patient is splinting (holding) his right side. There are coarse rhonchi (sounds that are evidence of secretions in the large airways of the lungs) and wheezes when you listen to the chest with your stethoscope. Blood tests reveal a P_aO_2 of 55 mm Hg (partial pressure of oxygen in arterial blood; normal > 90), a P_aCO_2 of 50 mm Hg (partial pressure of carbon dioxide in arterial blood; normal, 38–42), and a pH of 7.32 (normal, 7.38–7.42). The blood alcohol level is three times the legal limit. What part or parts of the respiratory system are not functioning correctly to account for the abnormal blood test results?

The patient has evidence of problems with the respiratory controller (alcohol is suppressing the part of the brain responsible for ventilation, as evidenced by the patient's slow breathing pattern and high P_aCO_2), the ventilatory pump (as evidenced by his fractured ribs, splinting, and shallow breathing), and gas exchanger (as evidenced by the fact that his P_aO_2 is lower than can be explained by the reduced ventilation alone). By the time you complete this book, you will be able to delineate the physiological principles underlying many of the findings on the patient's physical examination and describe the concepts that account for the gas exchange and acid–base abnormalities. Along the way, we hope that you will develop a strong appreciation, if not the passion, that we have for the beauty of the physiology of the respiratory system.

Summary Points

The text of each chapter ends with a list of Summary Points. These points provide a fairly complete review of the important points covered in the chapter. By reviewing these, along with the learning objectives outlined at the beginning of the chapter and the review questions at the end, you will be able to do a self-check of your understanding of the material presented.

- The study of the physiology of breathing requires an understanding that the achievement of respiration adequate to meet the metabolic needs of the body requires the coordinated interactions of the different elements of the respiratory system. One does not breathe with the lungs alone.
- The major components of the respiratory system are the respiratory controller, the ventilatory pump, and the gas exchanger.
- To master the material in this book and to be able to apply the concepts outlined, you should answer all thought questions and carefully analyze the Animated Figures.

Answers TO THOUGHT QUESTIONS

1-1. The study of the dynamics of the respiratory system is an analysis of its properties as the system is set in motion—that is, we are observing the system over time as air moves into and out of the lungs. The flow of gas, change in volume per unit time, is an important parameter as we study the system under dynamic conditions. One might surmise that the dynamic properties of the ventilatory pump would be affected at space travel near the speed of light. As one approximates the speed of light, time (as we understand it) changes. In contrast, under static conditions, we are examining the respiratory system when there is no flow of gas; thus, time is not a relevant variable, and space travel might be expected to have little effect on these properties.

Form and Function:
The Physiological Implications of the Anatomy of the Respiratory System

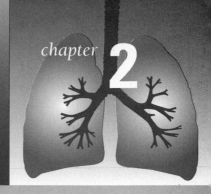

chapter **2**

CHAPTER OUTLINE

THE RESPIRATORY CONTROLLER
- Requirements of the Controller
- Automatic Control: The Central Pattern Generator
- Behavioral Control

THE VENTILATORY PUMP
- Requirements of the Ventilatory Pump
- Bones
- Muscles
- Pleura
- Peripheral Nerves
- Airways

THE GAS EXCHANGER
- Requirements of the Gas Exchanger
- Large Surface Area for Diffusion
- Minimizing Distance for Diffusion
- Stabilizing the Gas-exchanging Unit
- Matching Ventilation and Blood Flow
- Specialized Functions: Cells Lining the Alveoli

PUTTING IT TOGETHER

SUMMARY POINTS

LEARNING OBJECTIVES

- To define the key anatomic components of the controller.
- To delineate the functions, in addition to gas exchange, that the controller must be designed to address.
- To identify receptors, located throughout the body, that provide important information to the controller.
- To define the key anatomic components of the ventilatory pump.
- To describe the interplay between the muscles and the rib cage in the generation of a negative intrathoracic pressure during inspiration.
- To delineate the role of accessory muscles of ventilation in assisting the diaphragm during inspiration under conditions of high metabolic demand or abnormal chest wall function.
- To describe the role of the pleura in linking the motion of the chest wall and the lungs.
- To outline the peripheral nerves that are important in linking the controller and the ventilatory pump.
- To define the anatomic components of the gas exchanger.
- To describe the functional implications, particularly with regard to dead space and resistance, of the anatomy of the airways.
- To describe the functional implications, with particular regard to diffusion of gas, of the anatomic relationship between alveoli and capillaries.

9

- To delineate how the anatomic arrangement of alveoli contributes to their stability and minimizes the probability of alveolar collapse.
- To outline how the response of pulmonary arterioles to hypoxia improves gas exchange.

As a general principle, form and function are intimately related in the human body. Nature selects for features that provide the organism with a survival advantage. With this perspective, we may occasionally use teleological reasoning—that is, an analysis that views anatomic development of the respiratory system in the context of its function as we understand it today—to hypothesize links between anatomy and physiology. We consider the respiratory system in light of its major roles, not only with respect to gas exchange but also with an eye toward ways that the system supports other activities and assists in the maintenance of homeostasis.

The basic goal of the respiratory system is to transport oxygen from the atmosphere to the **alveoli**, the distal air sacs in the lung, where it can be picked up by hemoglobin in red blood cells (RBCs) and transported to metabolically active tissue. At the tissue level, oxygen is used within mitochondria to support aerobic metabolism, one byproduct of which is carbon dioxide. Carbon dioxide is then transported out of the body. However, the system must also serve as the "first responder" in the event of metabolic acid–base disorders. In addition, it must support speech, and the muscles of ventilation may, at times, be called upon to assist in other activities performed by the body.

The Respiratory Controller

The *respiratory controller* is the term used to denote the various elements of the system responsible for producing the neurological output of the brain that determines the rate and depth of breathing. Information is collected from a number of receptors and processed to lead ultimately to the neural information conducted by peripheral nerves to the ventilatory muscles (Fig. 2-1).

REQUIREMENTS OF THE CONTROLLER

To ensure adequate oxygen for metabolism 24 hours a day, including the hours that we are asleep, we need to have an automatic breathing center, meaning one that operates regardless of whether we are conscious and thinking about breathing. Imagine what would happen, however, if you were about to swallow some food and the diaphragm was suddenly stimulated to contract by the automatic control center, thereby generating a negative pressure in the thorax and movement of air into the lungs; food might well be sucked into the lungs as well. To avoid this misfortune and to allow us to speak, swim, and, on occasion, use our ventilatory muscles to support activities such as lifting heavy objects, we must also be able to modify or override the activity of the automatic center. Finally, the control system must be designed to monitor and respond to **acidemia**, a decrease in the normal pH of the blood.

QUICK CHECK 2-1	REQUIREMENTS OF THE RESPIRATORY CONTROLLER

Automatic control
Behavioral or voluntary control
Ability to monitor and respond to acidemia

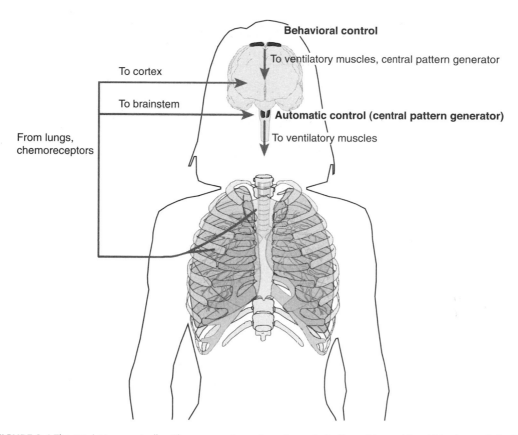

FIGURE 2-1 The respiratory controller. The neurons for automatic control of breathing reside within the medulla. This central pattern generator receives afferent information from the lungs and chemoreceptors and sends efferent information to the muscles of respiration via the spinal cord. Behavioral or conscious control of breathing originates in the motor cortex.

AUTOMATIC CONTROL: THE CENTRAL PATTERN GENERATOR

The primary source of automatic respiratory rhythm appears to reside with neurons located in the region of the brainstem called the **medulla**. These neurons have both inspiratory and expiratory activity, and we refer to this area of the brain as the **central pattern generator**. Within the medulla, the respiratory neurons appear to be collected into two regions: the dorsal respiratory group (DRG) and the ventral respiratory group (VRG). Information from receptors in the lungs and vascular system may be processed in the DRG. Most of the activity within the DRG, based on animal studies, is inspiratory in nature. The VRG contains both inspiratory and expiratory neurons, as well as cells that innervate muscles that control the larynx and pharynx. Axons from these neurons descend into the cervical spinal cord and ultimately project to the phrenic and intercostal nerves as well as accessory muscles, such as the sternocleidomastoid and the abdominal muscles. The accessory muscles are not primarily ventilatory muscles, but they do participate in ventilation when the body's metabolic needs are increased or if a mechanical problem exists with the ventilatory pump. As already noted, the motor neurons that activate the motion of the larynx are located within the VRG adjacent to inspiratory neurons. Thus, when these neurons send signals to initiate a breath, the muscles of the larynx are stimulated and the vocal cords abduct or move laterally outward, thereby decreasing the resistance of the upper airway and facilitating movement of air into the lungs.

TABLE 2-1 Airway and Pulmonary Receptors				
LOCATION	**MYELINATION**	**TYPE**	**STIMULUS**	**EFFECT ON VENTILATORY CONTROL**
UPPER AIRWAY				
Nose	Yes	Mechanical	Flow	Decrease ventilation
Pharynx	Yes	Mechanical	Swallow	Stop breathing
PULMONARY				
Slowly adapting receptors	Yes	Mechanical	Lung inflation	Prolong expiratory time Terminate inspiration
			Lung deflation	Increase respiratory rate
Rapidly adapting receptors	Yes	Mechanical and chemical	Lung deflation	Increase respiratory rate Sighs
C fibers	No	Mechanical	Increased pulmonary capillary pressure	?Increased respiratory rate
		Chemical	Capsaicin Bradykinin Serotonin Prostaglandin	?Increased respiratory rate

Respiratory related neurons have also been identified in the **pons**, the portion of the brainstem superior to the medulla, although their function is not well delineated. These neurons are collected in a structure called the pontine respiratory group (PRG). Researchers speculate that the PRG assists the body in making a smooth transition from inspiration to expiration during the respiratory cycle, and they may serve to coordinate information from higher centers with the activity of the central pattern generator. Injuries to the PRG, as may occur with a stroke, can lead to apneustic breathing, a respiratory pattern characterized by breaths with a very long inspiratory phase followed by a rapid exhalation and a brief pause before the next inspiration.

The central pattern generator, while creating the inherent rhythm of breathing, does not act in isolation. **Afferent information**, a term that denotes neurological sensory messages arising in peripheral nerves and transmitted to the central nervous system (CNS), comes from a variety of sources. Within the lungs, flow, pressure, stretch, and irritant receptors provide data on the movement of gas and distention of the lung (Table 2-1). Chapter 6 discusses the function of these receptors in more detail.

Several specialized sensory bodies called **chemoreceptors** monitor oxygen and carbon dioxide levels as well as the pH of the blood, and the chemoreceptors have fibers that extend to the medulla in the region where the inspiratory neurons are located. The peripheral chemoreceptors are located in the aortic arch and in the carotid bodies at the bifurcation of the common carotid artery (note that, in humans, the carotid chemoreceptors appear to have a much more important role than the aortic chemoreceptor). The **peripheral chemoreceptors** respond to hypoxemia (low partial pressure of oxygen in the blood), hypercapnia (elevated levels of carbon dioxide), and changes in pH in the blood. The **central chemoreceptors**, located in the brainstem, are sensitive to changes in arterial carbon dioxide levels and pH (Chapter 6 discusses the physiology of these structures more fully). You might view this arrangement as a way for the automatic center to monitor the activity of the entire respiratory system at any given time and to make adjustments accordingly without the need for conscious intervention.

BEHAVIORAL CONTROL

In contrast to the cardiovascular and gastrointestinal systems, which are under only automatic control, the respiratory system is also under conscious control. You can instantly double the size of a breath or the rate at which you are breathing. Alternatively, you can hold your breath. Imagine trying to talk, sing, or swim without the ability to exert some influence on the central pattern generator. Of course, the automatic center cannot be inhibited indefinitely; the gas exchange abnormalities associated with breathholding stimulate the chemoreceptors to the point that your ability to hold your breath is limited.

Volitional efforts to initiate or increase breathing are generated in the motor cortex and descend directly to the relevant ventilatory muscles; these signals are not processed through the respiratory centers in the medulla. The role of this pathway can be seen in individuals with a rare congenital abnormality, congenital central hypoventilation syndrome (CCHS). In this condition, which is estimated to affect approximately 200 individuals worldwide, the central pattern generator is inoperative; there is no automatic control of breathing. Despite the absence of a functioning central pattern generator, however, people with CCHS are able to voluntarily initiate breaths.

 THOUGHT QUESTION 2-1: What would you expect to happen to breathing in an individual with CCHS when she goes to sleep? Why?

Patients with CCHS provide an example of the independence of the automatic and volitional control of breathing. Breathholding offers an example of the integration of the two control mechanisms. Based on a unique experiment in which cats were trained to hold their breath, we believe that a breathhold is achieved by inhibiting the activity of the inspiratory neurons in the medulla (Orem J, Netick A: Behavioral control of breathing in the cat. *Brain Research* 1986, 366:238–253).

The Ventilatory Pump

The ventilatory pump is composed of the bones, muscles, and soft tissue of the thorax, the pleura lining the chest wall and the lungs, the peripheral nerves connecting the CNS to the ventilatory muscles, and the airways of the lung. All of these structures are necessary for the "bellows" function of the respiratory system, which is the movement of gas by bulk flow measured in liters per minute (in contrast to movement by diffusion measured in milliliters per minute) from the atmosphere to the alveoli and back out (note that we use the term "gas" as a generic term for the components of the atmosphere [primarily nitrogen and oxygen] and the contents of the alveoli [primarily nitrogen, oxygen, carbon dioxide, and water vapor]; there is a tendency to think of oxygen being inhaled and carbon dioxide being exhaled but, as you will see in Chapter 5, there are times when you must pay attention to the presence of these other gases as well).

REQUIREMENTS OF THE VENTILATORY PUMP

To move 5 L/min of gas into and out of the lungs, the ventilation level seen in a healthy, resting person, the ventilatory pump must create a negative pressure within the thorax (note: all pressures in respiratory physiology are taken relative to atmospheric pressure, which, by convention, is considered to be the zero reference point). In addition, the pump must provide a system for distribution of the inhaled gas to the alveoli. Finally, the pump must minimize energy expenditure to achieve its purpose while adapting to a range of required metabolic needs (e.g., exercise) and to diseases such as asthma that adversely affect airflow.

✔ QUICK CHECK 2-2   **REQUIREMENTS OF THE VENTILATORY PUMP**

Create negative intrathoracic pressure
Distribute gas throughout the lung
Minimize expenditure of energy while achieving adequate ventilation

As we examine the form and function of the ventilatory pump, we need to consider each of the key elements that contribute to the pump. These elements are the bones, muscles, pleura, peripheral nerves, and airways (Fig. 2-2).

BONES

The skeleton forms the superstructure on which muscles and connective tissue are placed. The internal organs of the thorax are encased and protected by the vertebral column, ribs, and sternum. In addition, these structures provide a relatively rigid cage that enables the muscles to generate a negative intrathoracic pressure during inspiration.

The ventilatory muscles, acting on the bones, expand the dimensions of the thorax. Viewed from the side of a person, the orientation of the ribs at the end of exhalation is "superior to inferior," that is, from the head downward toward the feet, as one goes from the vertebral column in the posterior region of the chest to the sternum anteriorly. During inspiration, the action of the external intercostal muscles lifts the ribs and makes them more horizontal. This movement has the effect of increasing the cross-sectional area of the thorax and contributes to the generation of negative intrathoracic pressure.

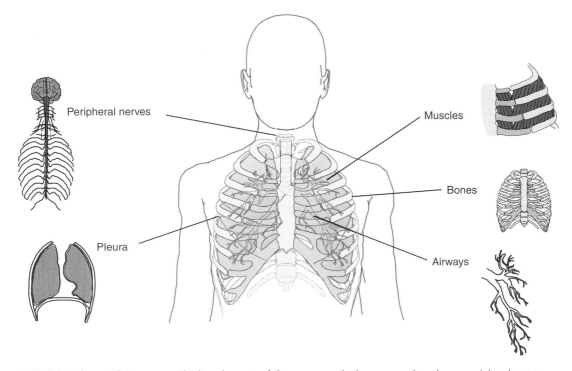

Peripheral nerves

Muscles

Bones

Pleura

Airways

FIGURE 2-2 The ventilatory pump. The key elements of the pump are the bones, muscles, pleura, peripheral nerves, and airways.

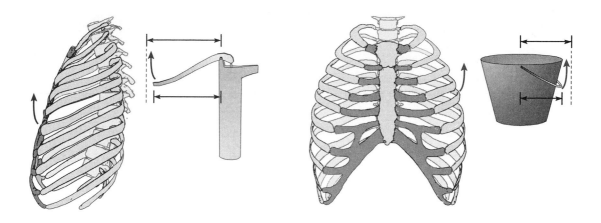

FIGURE 2-3 Rib motion during breathing. Note that the anatomic alignment of the ribs at the end of exhalation is superior to inferior as one moves from the vertebral column in the posterior portion of the chest to the sternum in the anterior thorax. As the external intercostal muscles contract, the ribs are lifted and become more horizontal, thereby increasing the cross-sectional area of the thorax. Seen from the frontal view, the lateral elevation of the ribs is analogous to the movement of a bucket handle. Seen from the side view, the elevation of the ribs and the superior motion of the sternum are similar to the action of a pump handle. Also see Animated Figure 2-3.

The sternocleidomastoid muscle, one of the accessory muscles of inhalation (see the section "Muscles of Inspiration"), inserts on the clavicle and sternum and helps lift the superior portions of the chest when greater force is needed to assist ventilation. These two motions of the rib cage—the upward lift of the ribs laterally and the sternum anteriorally—have been likened to the motion of bucket handle, in the case of the ribs, and the handle of an old-fashioned pump, in the case of the sternum (Fig. 2-3). Both movements work to increase the volume of the thoracic cavity.

Use Animated Figure 2-3 to view the pump handle and bucket handle motions of the ribs during breathing. You can press the play button to watch the motion during inspiration and expiration or use the slider to control the animation yourself.

MUSCLES

Inspiration, the movement of gas into the lungs, requires the generation of a negative pressure within the alveolus. Expiration, the movement of gas out of the lungs, requires positive pressure in the alveolus. The inspiratory muscles are necessary to initiate each breath. In normal individuals at rest, expiration is a passive phenomenon as the recoil of the lungs generates the pressure necessary to expel gas from the lungs. If airway resistance is high, as in asthma, or ventilatory requirements are increased, as during exercise, however, expiratory muscles must be activated.

Muscles of Inspiration

During quiet breathing at rest, inspiration is achieved largely as a consequence of the action of the diaphragm, with some assistance from the external intercostal muscles, located between the ribs, which lift and stabilize the rib cage. The diaphragm is innervated by cervical spinal roots that exit the spinal cord between C3 and C5. The intercostal muscles, in contrast, are activated by nerve roots that emanate from the spinal cord along the range of the thoracic vertebrae.

THOUGHT QUESTION 2-2: As you observe the chest wall during breathing, how would you predict the sternum would move during inspiration in a patient who has a complete transection of his spinal cord at the C7 level? Why?

Under conditions that require higher levels of ventilation or greater muscular force because of stiffness of the chest wall, obstruction of the airways, or weakness of the diaphragm, additional muscles, the accessory muscles of ventilation, must be used to assist in the generation of negative intrathoracic pressure. The major accessory muscles of inhalation include the scalenus and sternocleidomastoid muscles in the neck and the pectoralis muscles in the chest. For the pectoralis muscle, which normally acts to move the arm, to function as a ventilatory muscle, the arm must be fixed in position. Typically, patients in respiratory distress learn to do this on their own and assume a "tripod position" by resting their hands on their knees or on a nearby table or chair (Fig. 2-4). With the position of the arms secure, contraction of the pectoralis results in elevation of the anterior wall of the chest. Chances are you have also done this at some point after heavy exercise as you bent over with your hands on your knees and tried to catch your breath. A patient in the tripod position is likely to have a problem with the ventilatory pump or a need for high levels of ventilation.

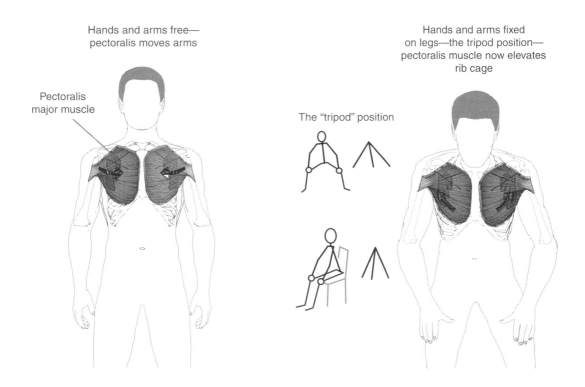

Hands and arms free—
pectoralis moves arms

Pectoralis
major muscle

Hands and arms fixed
on legs—the tripod position—
pectoralis muscle now elevates
rib cage

The "tripod" position

FIGURE 2-4 The tripod position. With the hands free, the pectoralis muscles act to move the arms. Notice that when the patient puts his hands on his knees, this position fixes the arms and prevents them from moving. Activation of the pectoralis muscles now results in the elevation of the rib cage. The function of the pectoralis muscle has been transformed. It now serves as an accessory muscle of ventilation.

Muscles of Expiration

During quiet breathing at rest, expiration is a passive process. When the inspiratory muscles relax at the end of inspiration, the normal tendency of the lungs to recoil inward, which is the consequence of elastic and surface forces (Chapter 3 discusses the exact nature of these forces), produces positive pressure within the lung. A pressure gradient is thus established between the alveolus and the mouth. Consequently, air moves out of the thorax. In a sense, the energy stored in the lungs by the action of the inspiratory muscles is now used for expiration, a system design that minimizes energy expenditure.

To increase ventilation, which is the volume of air going into and out of the lungs each minute, one can increase **tidal volume** (the size of the breath) and/or the rate or frequency of breathing. For small to moderate increases in ventilation, changes in tidal volume may suffice. Above a ventilation of about 30 to 40 L/min, however, the rate of breathing must also be increased. To increase the breathing rate, you must reduce the amount of time that the system spends in exhalation (i.e., you must increase expiratory flow). Because expiration is passive during normal, quiet breathing, expiratory flow is determined under resting conditions by the pressure in the alveolus, which is produced by the recoil of the lungs, and the resistance of the airways (see Chapter 4). Therefore, to increase expiratory flow, you must increase the pressure gradient between the alveolus and the pharynx by recruiting accessory muscles of exhalation. The major accessory muscles of exhalation are the internal intercostal and the abdominal muscles. The nerves that innervate the intercostal and abdominal muscles originate from the thoracic and lumbar spine.

PLEURA

The lungs are surrounded by a thin layer of tissue called the **visceral pleura**. A similar layer, called the **parietal pleura**, lines the inside of the chest wall. The space between these two layers, the **pleural space**, is a virtual space containing only a few milliliters of fluid. Physiologically, the pleurae and the space they create play a critical role in linking the motion of the chest wall and the lungs. The chest wall and the lungs possess elastic properties that vary over the course of the breath. At the end of exhalation, the elastic properties of the lung exert a force inward (i.e., a collapsing force); at the same volume, because the resting volume of the chest wall is greater than the volume at the end of exhalation, the elastic properties of the chest wall exert a force in an outward direction (i.e., an expanding force). At this point in the respiratory cycle, these forces are equal and opposite to each other and are linked via the pleural space. The outward recoil of the chest wall and the collapsing forces of the lung create a negative pressure within the pleural space. When the chest wall moves outward because of the action of the inspiratory muscles, the pressure within the pleural space decreases (i.e., becomes more negative), thereby leading to a decrease in alveolar pressure and flow of air into the lung. When the inspiratory muscles relax at the end of inspiration, the pleural pressure increases, or becomes less negative, and the elastic forces of the lung produce a positive alveolar pressure and flow of gas out of the lung. This interaction and sequence of these events are explained in more detail in Chapter 3.

PERIPHERAL NERVES

Motor nerves emanate from the spinal cord and link the controller to the muscles of ventilation. As already discussed, the diaphragm receives its innervation from cervical nerve roots (C3–C5), and the intercostal muscles are activated via nerves from the thoracic spinal cord. The abdominal muscles, which are the strongest expiratory muscles, are innervated by nerves from the thoracic and lumbar spine (Fig. 2-5).

The lungs are rich in sensory receptors that detect changes in flow and pressure in the airways as well as volume changes in the lung parenchyma. In addition, C fibers in airways and near pulmonary capillaries may play a role in sensations associated with the buildup of fluid in the lung, as is seen in heart failure. One could argue that the lungs are, in many ways, sensory organs (see Chapter 8). All of the information arising from these receptors is conveyed to the brain via the **vagus nerve**.

Finally, the autonomic nervous system also plays a role in regulating the ventilatory pump. Fibers from the **sympathetic nervous system** terminate near the airways. When stimulated, these fibers lead to bronchodilation. Conversely, stimulation of the **parasympathetic nervous system**, which innervates the airways via cholinergic receptors, causes

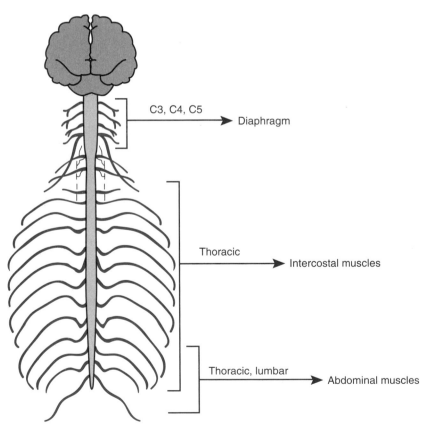

FIGURE 2-5 Peripheral nerves to the muscles of ventilation. The cervical nerve roots (C3–C5) innervate the diaphragm, and the thoracic spinal cord innervates the intercostal muscles. Neural projections to the abdominal muscles come from the thoracic and lumbar spine.

bronchoconstriction. The balance between these two components of the autonomic nervous system determines the tone of the muscles of the airways and the resulting diameter of these conducting tubes (and resistance to air flow within them).

> **?**
>
> **THOUGHT QUESTION 2-3:** You are evaluating a patient who experiences shortness of breath during running. You are concerned that he may have asthma. His lung function at rest is normal. If you wanted to test whether this patient has asthma (an increased sensitivity to develop constriction of the airways), would you administer to the patient a β-agonist or a cholinergic agonist? Why?

AIRWAYS

The airways resemble the branches of a tree that ultimately end in the leaves (alveoli) where gas exchange takes place. The airways serve as the conduit for the flow of gas from the mouth to the alveoli and back out.

Every time an airway branches and a single airway becomes two, we say we have moved to another "generation" of airways. The single trachea, or windpipe, bifurcates into two large tubes that lead to the left and right lungs and are called *mainstem bronchi*. The mainstem bronchi divide into smaller tubes, called *lobar bronchi*, which lead to the lobes of the lungs. The right lung has three lobes, and the left lung has two. A lobe is completely surrounded by pleura, so air cannot move from one lobe to another except through the airways. Within each lobe are segments, which receive air from branches of the lobar bronchi called *segmental bronchi*. Segments are divided into subsegments and receive air via subsegmental bronchi, which lead ultimately to the **terminal bronchioles**, the smallest airways before the alveoli appear. Bronchi contain cartilage, submucosal glands, ciliated epithelial cells, and goblet cells. The ciliated cells and goblet cells form part of the lungs' defense system. Goblet cells produce mucus. Foreign material and infectious agents may be trapped in the airway mucus, which is then moved toward the pharynx by the cilia, hairlike structures that have a rhythmic beat. When it reaches the pharynx, the mucus is either swallowed or coughed out. The bronchioles do not normally contain cartilage, glands, or goblet cells.

The sum of the cross-sectional area of the two new airways at each branch point is greater than that of the parent branch. Thus, as you move farther out toward the alveoli, the cross-sectional area of all of the small airways in parallel is quite large, far larger than the trachea (Fig. 2-6).

The number of branch points from the trachea to the alveolus ranges from 10 to 23. As you move from **conducting airways**, tubes through which air flows but no gas exchange takes place, to the alveoli or gas exchange units, you go through what is known as the **transitional zone**. This zone consists of airways called **respiratory bronchioles**, from which a few alveoli originate; thus, there is a transition between the conducting airways and alveoli. Finally, you arrive at the alveolar ducts, which are completely lined by alveoli and are the primary site of gas exchange. The portion of the lung composed of alveoli has been termed the **respiratory zone**.

> **?**
>
> **THOUGHT QUESTION 2-4:** Where would you expect to find the highest resistance to airflow in the lung: in the large central airways or in the very tiny, peripheral airways? Why?

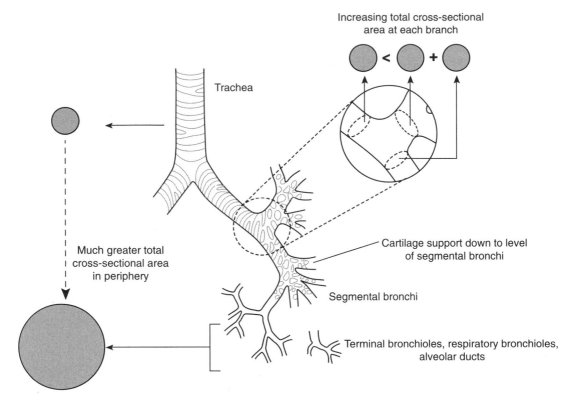

FIGURE 2-6 Airways of the lungs. As you travel from the mouth to the alveoli, you encounter an increasing number of branch points. As a general principle, at each branch point, the sum of the cross-sectional areas of the new airways is greater than that of the airway from which they originated. The more central airways have cartilaginous support, resulting in some stiffness in their walls. However, the smaller, distal airways are quite compliant and prone to collapse when the pressure outside the airways is greater than the pressure inside.

QUICK CHECK 2-4	THE MANY GENERATIONS OF AIRWAYS
Conducting Airways • **Trachea** • **Mainstem bronchi** • **Lobar bronchi** • **Segmental bronchi** • **Subsegmental bronchi** • **Terminal bronchioles**	**Transitional Zone** • **Respiratory bronchioles** **Gas-exchanging Zone** • **Alveolar ducts**

Regions of lung that receive air but do not participate in gas exchange are termed **dead space**. The size of the conducting airways offers a tradeoff between two competing physiologic demands. On the one hand, the amount of wasted ventilation or dead space needs to be minimized. This would favor conducting airways with very small diameter and volume. On the other hand, the resistance in these airways must also be minimized. Because airway resistance varies inversely with the radius raised to the fourth power (see Chapter 4), the need to minimize the work of breathing suggests that the system should have the largest airway possible. Hence, you can see the dilemma.

Moving from the single trachea to millions of respiratory bronchioles arranged in parallel, the cross-sectional area of the airways increases tremendously. Consequently, the velocity of the air moving through the airways decreases as it travels from the trachea to the periphery of the lung. Consider the rapidly moving water in a creek that empties into a large pond; the velocity of the water decreases markedly as it enters the broad pond. Because the total cross-sectional area is a key factor in determining resistance, defined as a pressure decrease divided by flow, most of the resistance in the lungs is in the central airways. The first six branches of the airways dominate the total resistance of the lung.

The bronchioles are generally less than 1 mm in diameter. As already noted, the bronchioles do not have cartilage in their walls. Thus, they are susceptible to collapse if the pressure in the pleural space is greater than the pressure in the airway. The bronchioles are located within the connective tissue structure of the lung and are supported by this tissue. As the lung increases in volume during inspiration, the diameter of the bronchioles increases as well. As the lung volume decreases during expiration, so does the diameter of the bronchioles. Emphysema, a disease associated with cigarette smoking, destroys much of the connective tissue support structure within the lung, thereby making these small airways even more susceptible to collapse.

Gas moves from the pharynx to the terminal bronchioles by bulk flow. The driving force during inspiration is the pressure differential between the pharynx and the alveolus (alveolar pressure is negative during inspiration because of the action of the inspiratory muscles). Beyond the terminal bronchioles, the cross-sectional area of the millions of respiratory bronchioles and alveolar ducts, all arranged in parallel, becomes enormous and the velocity of the gas entering the gas-exchanging units decreases precipitously (see Chapter 4). In this region of the airways, diffusion is the predominant means for moving gas. The very low velocity of gas in this region of the lung allows particles being carried in the inspired gas to settle out. The low velocity here likely accounts for the fact that this anatomic region is often the site of deposition of very small dust particles (and the site of diseases that are associated with these particles) (Fig. 2-7).

Because the respiratory bronchioles and alveolar ducts contribute little to the overall resistance of the lungs, the distribution of gas among these units is determined largely by their relative **compliance**. Compliance is a measure of the stiffness of an object and is equal to the change in volume that occurs in the object for a given change in the pressure across the wall of the object.

$$\text{Compliance} = \frac{\Delta \text{ Volume}}{\Delta \text{ Pressure}}$$

The more compliant units receive more ventilation. In the normal lung, the compliance of a terminal lung unit depends on the volume of the unit—the larger the unit at the beginning of a breath, the less compliant it is (see Chapter 3). At the end of exhalation, alveolar units at the apex of the lung are larger than at the bases and, as a result, are less compliant. This accounts partly for the finding that inspired gas is distributed preferentially to the bases of the lungs relative to the apices (see Chapter 5). The distribution of ventilation within the lungs is also affected by microscopic passages called the **pores of Kohn**, which connect the alveoli within a lobe (Fig. 2-8). These connections permit transfer of gas between alveoli and function to minimize the collapse of lung units if a more central airway is obstructed. As noted previously, however, individual lobes of the lung are fully surrounded by pleura, and cross-ventilation between the lobes can only occur via the airways.

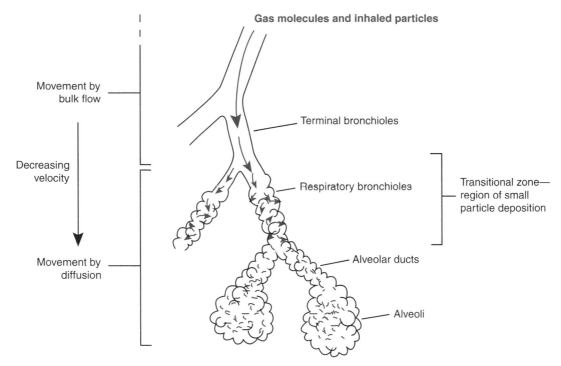

FIGURE 2-7 The transitional zone of the lung. As gas moves from the terminal bronchioles to the respiratory bronchioles and the alveolar ducts, the velocity of the gas decreases markedly. Gas now moves largely by diffusion rather than bulk flow. The decrease in velocity leads to deposition of small, inhaled particles in this region of the lung.

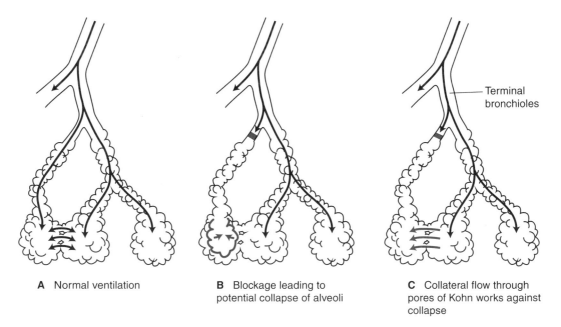

FIGURE 2-8 Pores of Kohn. Connections between alveoli, called pores of Kohn, allow gas transfer between alveoli within a pulmonary segment. This minimizes the chances that an obstruction within an airway will lead to collapse of alveoli and loss of gas exchanging units. **A,** The normal flow of gas from the respiratory bronchioles to the alveoli. **B,** An obstruction forms in a small airway leading to a group of alveoli. **C,** The movement of gas into these alveoli from adjacent units via the pores of Kohn prevents collapse of the alveoli downstream from the obstruction.

THOUGHT QUESTION 2-5: Immotile cilia syndrome is a genetically transmitted disorder characterized by cilia that do not move normally. Patients with this syndrome may be infertile (cilia line the fallopian tubes, and spermatic function is also affected) and typically have respiratory disease. Approximately half of the patients have situs inversus (the side of the body in which the internal organs are situated is reversed because the embryonic cilia responsible for pushing the organs to the correct place during fetal development do not function properly), in which case they are said to have Kartagener's syndrome. What parts of the lungs do you think would be affected in this condition? What types of problems might they have? Why?

In disease states such as chronic bronchitis, excess production of mucus may reduce the functional diameter of the airways and increase resistance to airflow.

The Gas Exchanger

If the controller and ventilatory pump have accomplished their respective tasks, air arrives at the gas exchanger, which is composed of the alveoli and the pulmonary capillaries. The bulk flow seen in the airways gives way to diffusion as oxygen leaves the alveolus and enters the blood and carbon dioxide moves in the opposite direction. As with the other components of the respiratory system, nature has constructed a very efficient structure to perform this vital function.

REQUIREMENTS OF THE GAS EXCHANGER

Having moved air from the pharynx to the respiratory bronchioles, the system must now provide a large surface area for the diffusion of oxygen into and carbon dioxide out of the blood. In addition, to ensure efficient transfer of gas between the blood and the alveolus, the distance for diffusion must be as short as possible. The gas exchanging unit, however, must be sufficiently robust to resist collapse during exhalation as the lung volume decreases. Finally, the system must ensure that **perfusion**, the term given to blood flow to an organ or tissue bed, goes to the areas of the lung where oxygen is being delivered.

QUICK CHECK 2-5   **REQUIREMENTS OF THE GAS EXCHANGER**

Provide large surface area for the diffusion of gas
Minimize distance for diffusion
Maintain sufficient structural integrity of the gas-exchanging unit to minimize
   probability that the unit will collapse at low lung volumes
Match ventilation and blood flow

LARGE SURFACE AREA FOR DIFFUSION

The amount of gas that can diffuse across a surface is proportional to the area of the surface. To maximize the surface area for gas exchange, millions of gas exchanging units are present at the ends of the respiratory bronchioles. These gas-exchanging units, or alveoli, can be found primarily as part of the alveolar ducts. A respiratory bronchiole, which also

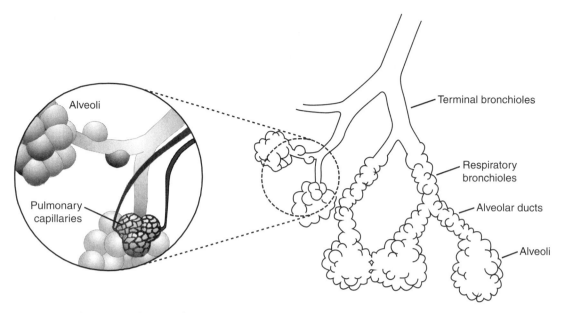

FIGURE 2-9 Pulmonary capillaries and alveoli. The pulmonary capillaries form a fine mesh network around each alveolus to maximize the contact between the two structures and, consequently, the surface area for diffusion of gas.

contains a few alveoli emanating from its wall, and its associated alveolar ducts are the site of gas exchange and constitute an **acinus** or **pulmonary lobule**. Roughly 300 million alveoli are present in the lung, each of which is less than 0.5 mm in diameter. The total surface area of the gas-exchanging units in the lungs is between 50 and 100 m^2.

Because the surface area of the alveoli is large, a comparably large pulmonary capillary surface area is needed to achieve effective transfer of oxygen and carbon dioxide between the two structures. The pulmonary capillaries are literally wrapped around the alveolus in a fine mesh network that produces a virtual sheet of blood for diffusion (Fig. 2-9).

As many as 1000 pulmonary capillaries come into contact with each alveolus. The diameter of a capillary is only 10 μm, a size that permits just a single RBC to pass through the vessel. The small diameter ensures that the gas has only a minimal distance to travel between the alveolus and the hemoglobin contained within the RBC. Hemoglobin is the primary protein within the RBC and, by virtue of its capacity to bind oxygen, is the major means by which oxygen is transported in the blood. Each molecule of hemoglobin contains four binding sites for oxygen, and we characterize the extent to which the blood is "filled" with oxygen by measuring the percentage of binding sites that are bound with the molecule. This percentage is called the **oxygen saturation**.

During the time that an RBC traverses a pulmonary capillary, it may actually come into contact with several alveoli. Assuming a normal cardiac output in a resting individual, the hemoglobin in an RBC is typically fully saturated with oxygen by the time it has made it 25% to 33% of the way through the pulmonary capillary. This means that there is a tremendous reserve capacity for diffusion. For example, when cardiac output is increased, as in exercise, and the speed with which an RBC traverses the alveolus increases, the blood will still be saturated with oxygen by the time it exits the capillary. Alternatively, if a disease process results in a thickening of the alveolar–capillary interface, there may still be adequate time for the oxygen to diffuse into the RBCs.

THOUGHT QUESTION 2-6: An individual with pulmonary fibrosis, a condition that results in scarring in the lung and thickening of the alveolar–capillary interface, comes to see you for evaluation of her oxygen saturation level. At rest, the patient's oxygen level is normal. What would you expect to happen to the oxygen level in the blood when the patient exercises? Why?

Under normal conditions, the walls of the alveolus and the capillary are so thin that the pressure within the alveolus can affect blood flow within the capillary. In some regions of the lung, pressure in the alveolus exceeds that within the capillary, leading to compression or collapse of the capillary (more in Chapter 5).

MINIMIZING DISTANCE FOR DIFFUSION

The amount of gas that can diffuse across a surface is inversely proportional to the thickness of the surface. The total distance for diffusion of gas from the alveolus to the RBC is approximately 0.5 μm. The surface across which the gas must diffuse consists of the fluid lining the alveolus, the alveolar wall, the interstitial space between the alveolus and the capillary (a space that is negligible in normal individuals but where fluid and inflammatory material may accumulate in disease states), the wall of the capillary, the plasma surrounding the RBC, and the wall of the RBC.

QUICK CHECK 2-6	PATH OF OXYGEN FROM THE ALVEOLUS TO HEMOGLOBIN IN THE RED BLOOD CELLS	
1. Fluid lining the alveolus	4. Capillary wall	
2. Alveolar wall	5. Plasma surrounding the RBC	
3. Interstitial space	6. Wall of RBC	

The wall of the pulmonary capillary is composed of a single layer of squamous epithelium, which minimizes the distance for diffusion. The wall of the alveolus is similarly thin, composed of an epithelium made up of a single layer of cells called type I pneumocytes and an interstitium formed by the fused basal lamina of the alveolar epithelium and the capillary endothelium (in some regions, collagen and elastin also contribute to the thickness of the interstitial space). In healthy people, the thickness of the RBCs forms a significant portion of the distance across which oxygen and carbon dioxide must diffuse.

Diffusion is so effective in the lungs, and the reserve time available for diffusion of oxygen into the RBCs is so great, that a low blood oxygen level attributable to diffusion limitation is difficult to demonstrate in the normal lung. In experimental conditions, a low blood oxygen level may occur in a person doing heavy exercise while breathing a mixture of gas with a low partial pressure of oxygen. Even in disease states, diffusion limitation is rarely a cause of **hypoxemia**, the term used for low blood oxygen levels at rest (see Chapter 5).

STABILIZING THE GAS-EXCHANGING UNIT

As the volume of alveoli decreases during expiration, surface forces and the physics of small, gas-containing objects increase the probability that the alveoli will collapse (see Chapter 3). Anatomically, the alveoli are linked in ways that minimize the probability of

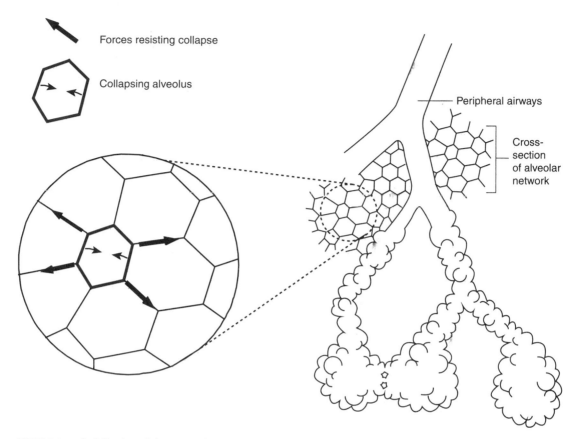

FIGURE 2-10 Stabilization of the gas-exchanging unit. Alveoli are linked to each other via shared walls and to airways via collagen and elastin. These connections allow forces to be shared among alveoli, and the resulting stabilization makes it difficult for a single alveolus to collapse. This form of structural support is also known as "interdependence." **Also see Animated Figure 2-10.**

collapse. Connective tissue, composed of collagen and elastin, is interspersed between the alveoli and the airways. As a consequence of this "connectedness," forces imposed at the level of the alveolar wall are transmitted to adjacent alveoli and the rest of the lung, thereby providing a form of stabilization to individual gas-exchanging units (Fig. 2-10).

Use Animated Figure 2-10 to view how the collapse of a single alveolus is opposed by its connections to surrounding alveoli. Drag one of the alveolar walls and release it to observe how the resulting forces return the collapsing alveolus to its initial state.

The wall of one alveolus is often shared by another alveolus. If one alveolus were to collapse, therefore, others would be affected as the adjacent alveoli are pulled down in the same direction. The phenomenon of **atelectasis**, or collapse of lung units, is not caused by collapse of individual alveoli; larger lung units must be involved given this degree of interdependence.

MATCHING VENTILATION AND BLOOD FLOW

For gas exchange to be optimized, the body must match the distribution of ventilation, or movement of air into the lung, with perfusion, the blood flow in the pulmonary capillaries. The pulmonary arteries are located next to the bronchi and branch with them down to the level of the respiratory bronchiole. Here, the arteries divide into capillaries that surround the alveoli, as already described. The pulmonary arteries, however, are not passive

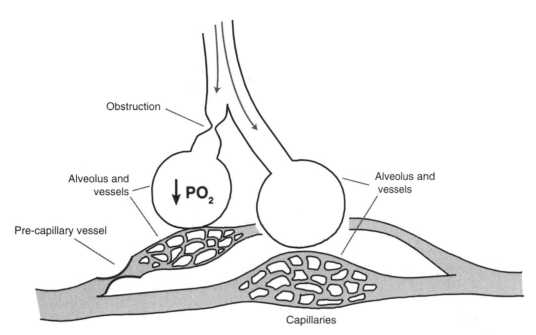

FIGURE 2-11 Hypoxic pulmonary vasoconstriction. Arterial smooth muscle cells in pulmonary arterioles respond to hypoxia by contracting. As ventilation decreases to one alveolus, the partial pressure of oxygen diminishes within that unit. This process leads to an increase in vascular resistance in the vessels perfusing the alveolus and a redirection of blood flow to other regions of lung that are better ventilated and have higher oxygen levels. This phenomenon maximizes gas exchange by matching ventilation and perfusion within the lung. **Also see Animated Figure 2-11.**

conduits for blood flow. The arterial walls contain smooth muscle that can contract, thereby changing the resistance of the vessel and the amount of blood flowing through a particular region of lung. Hypoxia is one of the stimuli that cause pulmonary arterial vasoconstriction. Thus, if one section of lung is not receiving adequate ventilation and local hypoxia develops in that area, the pulmonary artery serving that region will constrict to redirect blood flow to another area that is receiving adequate ventilation (Fig. 2-11). This phenomenon, termed **hypoxic pulmonary vasoconstriction**, ensures that blood flows to regions of lung receiving the best ventilation. Use Animated Figure 2-11 to adjust the airway obstruction and hence the ventilation to the alveolus. Note that as alveolar hypoxia develops (i.e., its partial pressure of oxygen, or PO_2, decreases), constriction of the pre-capillary vessel supplying the affected alveolus occurs, and blood flow is redirected to unaffected alveoli. By matching ventilation and perfusion, the body maximizes the opportunity for diffusion of oxygen and carbon dioxide. Mismatch of ventilation (represented by a V with a dot over it, $\dot{V}$) and perfusion (represented by the letter Q), termed $\dot{V}/Q$ mismatch, is the most common physiological cause of hypoxemia in patients with cardiopulmonary disease (see Chapter 5).

The pulmonary circulation must also be able to handle large changes in blood flow without causing a significant increase in vascular resistance. When an individual is resting, the volume of blood in the pulmonary capillaries is only one third of the capacity of the vessels, meaning there is a significant reserve capacity. When cardiac output increases, as during exercise, to meet the metabolic needs of the body, the amount of blood engaging in gas exchange can increase as well. As more blood enters pulmonary capillaries, additional alveoli are *recruited* into the gas-exchanging process, so blood flows to alveoli that previously received no blood, and $\dot{V}/Q$ matching is improved.

The lung has two separate circulations. Thus far, we have focused on the **pulmonary circulation**, which delivers deoxygenated blood from the right ventricle of the heart to the pulmonary capillaries, where oxygen is transferred to the RBCs and carbon dioxide is transferred to the alveoli. The **bronchial circulation** delivers oxygenated blood from the aorta to the lung tissue. The bronchial circulation serves as a major supply of oxygen and nutrients to the trachea, bronchi, and visceral pleura (as well as the esophagus). Together, these structures comprise the extra-alveolar components of the lungs. Bronchial blood flow may be greatly increased in inflammatory conditions of the lung such as bronchiectasis. Erosion of these vessels can lead to intrapulmonary bleeding and hemoptysis (i.e., coughing up blood).

THOUGHT QUESTION 2-7: A pulmonary embolism is a blood clot that travels from one of the veins in the body to the lungs and becomes lodged within the pulmonary circulation, thus blocking blood flow to that region of lung. Why are most pulmonary emboli not associated with pulmonary infarction (death of lung tissue)?

SPECIALIZED FUNCTIONS: CELLS LINING THE ALVEOLI

The alveoli are lined by cells that perform a number of important functions. Type I pneumocytes line the alveolus. Interspersed among these type I pneumocytes are other cells called **type II pneumocytes**, which produce **surfactant**, a substance that becomes a component of the liquid layer lining the alveolus. Surfactant plays a critical role in reducing surface forces within the alveolus, thereby contributing to the stability of the alveoli (see Chapter 3). Type II pneumocytes also absorb and recycle surfactant.

We have previously described the role of the airways in the body's defenses. Mucus traps inhaled foreign material, and the cilia push this material up to the trachea, where it can be expelled. The alveoli also play a part in defending the lungs from infection. Macrophages are found within the alveoli and ingest foreign material that has evaded the airway defenses and managed to enter the gas-exchanging units. These cells help police the alveoli against infectious agents to minimize damage to the lungs and prevent entry of these organisms into the bloodstream, where they can be carried to the rest of the body.

PUTTING IT TOGETHER

A 16-year-old boy with a history of asthma visits his friend, who recently received a pet cat. The boy with asthma is allergic to cats, and over the course of an hour, he notes that his breathing is becoming uncomfortable. He feels tightness in his chest and begins to cough. He has forgotten his medications and decides to "tough it out." Over the next hour, however, his breathing becomes increasingly difficult, and his friend's mother takes him to the emergency department.

The physician notes that the boy's respiratory rate is increased (respiratory frequency is 20/min), and the boy is contracting his sternocleidomastoid muscles with each breath. Diffuse wheezing is noted when the physician listens to the patient's lungs with his stethoscope. The oxygen saturation of the blood is only slightly down despite the wheezing. The doctor administers albuterol, a β-agonist that binds to the β-2 receptors on the airway smooth muscle, by inhalation, and the boy's symptoms improve considerably. What links can you make between the structure and function of the respiratory system in your analysis of this case?

Let's consider how the structure of the respiratory system worked to help sustain the boy during this acute asthma attack. Exposure to the cat dander, an allergen to which the boy was sensitive, led to constriction of the bronchioles and increased production of mucus within the airways. Both of these factors caused airway resistance to increase and diminished ventilation to some regions of the lung. The respiratory controller sensed changes in the lung via pulmonary receptors and the development of mild hypoxia via the chemoreceptors, as well as the distress the boy was feeling. The result was an increase in the rate of breathing. To deal with the impairment to the ventilatory pump caused by the constriction of the airways, accessory muscles of ventilation, the sternocleidomastoid muscles, were recruited to serve as ventilatory muscles. The oxygen level was only minimally reduced despite the abnormalities in the airways because the pulmonary capillaries supplying blood to regions of lung most impaired by the bronchoconstriction and mucus constricted in response to localized hypoxia. Thus, blood was sent to relatively unaffected areas of the lungs, and gas exchange was protected. Finally, the doctor administered a β-agonist to bind to β-2 receptors on the airway smooth muscle cells, thereby leading to bronchodilation.

Summary Points

- The structural anatomy of the respiratory system is intimately linked to the physiologic requirements of breathing and gas exchange.
- The respiratory controller has both automatic and voluntary elements. Consciousness is not required to keep the system in motion, but behavioral factors can modify the rate and depth of breathing.
- Feedback from a variety of receptors in the respiratory system helps modulate the control of breathing.
- The vertebral column and the ribs support the muscles of ventilation and protect the lungs and other intrathoracic organs.
- The action of the muscles of ventilation is important in the development of negative and positive pressure within the thorax. The accessory muscles normally serve other functions but are recruited to assist the ventilatory pump when the body needs to increase the rate and depth of breathing or when the ventilatory pump is impaired.
- The pleurae link the motion of the chest wall and the lungs.
- The peripheral and autonomic nervous systems help translate activity in the controller into motion of the ventilatory pump, provide a means for the controller to assess the status of the pump, and play a role in establishing the diameter of the airways.
- The airways provide a conduit for the bulk flow of gas from the pharynx to the alveoli and back out again. They also serve an important role in defending the body against infection.
- The size and distribution of the airways reflect the need to balance airway resistance and dead space.
- Gas flow in the respiratory bronchioles and alveolar ducts is determined by diffusion rather than bulk flow.
- The gas exchanger is designed to maximize the area for diffusion of gases and minimize the distance between the alveolar lumen and the hemoglobin within the erythrocyte.
- The pulmonary circulation is a low-resistance system that has the capacity to hold much more blood than it contains when the body is resting.
- Pores of Kohn and the structural interconnectedness of alveoli contribute to the stability of the gas-exchanging units.
- Hypoxic pulmonary vasoconstriction in the lungs is an important mechanism for matching blood flow and ventilation.
- Type II pneumocytes in the alveoli produce surfactant, a substance that plays an important role in reducing surface tension and preventing alveolar collapse.

Answers TO THOUGHT QUESTIONS

2-1.   Individuals with CCHS do not have a working central pattern generator. Consequently, when they sleep and all conscious activity ceases, they stop breathing. The condition is generally diagnosed shortly after birth because the infant will have a respiratory arrest when asleep. These people need a permanent tracheostomy, a hole in the trachea that permits the patient to be connected to a mechanical ventilator via a tube placed through the hole. The ventilator must be used to sustain them during sleep.

2-2.   Because the diaphragm receives its neurological signals from nerves that exit the spinal cord between C3 and C5, above the level of the injury, the diaphragm will contract and move normally, leading to the creation of negative intrathoracic pressure (see figure below). The intercostal muscles, however, derive their neurological signals from the thoracic segments of the spinal cord, so these muscles are inactive. Without the action of the intercostal muscles to stabilize the rib cage, the sternum will move inward during inspiration. In essence, the sternum is sucked in by the negative intrathoracic pressure.

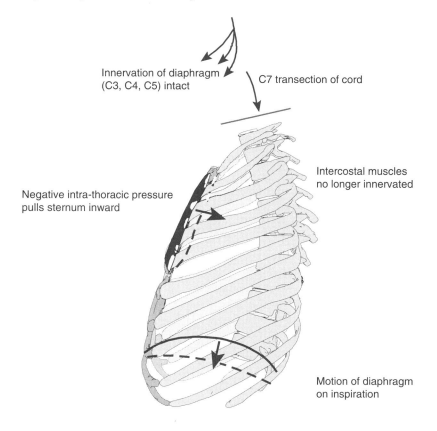

Innervation of diaphragm
(C3, C4, C5) intact

C7 transection of cord

Intercostal muscles
no longer innervated

Negative intra-thoracic pressure
pulls sternum inward

Motion of diaphragm
on inspiration

Lateral view of rib cage

With a complete transection of the spinal cord at C7, the diaphragm is still innervated and moves downward during inspiration. Without the action of the intercostal muscles, however, the negative intrathoracic pressure generated by diaphragm causes the sternum to move inward.

2-3.   A β-agonist would stimulate receptors in the sympathetic nervous system and lead to bronchodilation. A cholinergic agonist would stimulate the parasympathetic receptors and cause bronchoconstriction. A common way to test for the presence of asthma is to ask the patient to inhale gradually increasing doses of methacholine, a cholinergic agonist. A diagnosis of asthma is made if the patient develops significant bronchoconstriction at low doses of the inhaled agent.

2-4.   The highest resistance to airflow is found toward the large central airways. The great cross-sectional area of the millions of peripheral airways in parallel leads to low resistance in the periphery of the lung. As discussed in Chapter 4, one of the major factors in determining the resistance of the airways (expressed as the decrease in pressure divided by the flow) is the presence or absence of turbulent flow, which depends largely on the velocity of the gas moving through the airway. As gas moves from a single windpipe to millions of terminal bronchioles, the velocity of the gas decreases tremendously (think of water swiftly moving through a creek that empties into a large pond: the velocity of the water slows dramatically as it enters the pond), and resistance falls. It is estimated that only 20% of the total resistance in the lungs is attributed to airways less than 2 mm in diameter.

2-5.   The airways from the trachea to the respiratory bronchioles are lined with cilia. If the cilia do not function normally, the mucus and the foreign material trapped in the mucus are not pushed back up to the trachea to be either swallowed or coughed out. This abnormality leads to recurrent respiratory infections because one of the key defense mechanisms that prevents bacteria from entering the lungs is disrupted. Recurrent infections and the inflammatory response to them can, over time, lead to dilatation and scarring of the airways, a condition called bronchiectasis.

2-6.   In the patient described, the oxygen level in the blood will decrease when she exercises. The distance for diffusion has been increased because of the scarring of the lung. Although the reserve capacity of the gas-exchanging unit can compensate for this when the patient is at rest, the increase in cardiac output associated with exercise puts a further strain on the system. The RBC is now speeding through the pulmonary capillary, and the prolonged time required for diffusion, consequent to the pulmonary fibrosis, prevents the hemoglobin from being fully saturated as the RBC exits the capillary. As a result, the blood oxygen level decreases with exercise.

2-7.   The lung tissue can receive oxygen from multiple sources: the airways, the bronchial circulation, and the pulmonary circulation. Thus, a blockage of one component of the pulmonary circulation does not necessarily lead to sufficient hypoxia to cause cell death.

chapter

2

Review Questions

DIRECTIONS: *Each of the numbered items or incomplete statements in this section is followed by answers or by completions of the statement. Select the ONE lettered answer or completion that is BEST in each case.*

1. A 22-year-old college student falls into a pool after consuming large amounts of alcohol. He is unable to swim and sinks to the bottom of the pool. An alert friend walks by approximately 4 minutes later and dives in after him. He pulls his friend to the surface and begins cardiopulmonary resuscitation. The student is revived and is taken to the hospital, where he is ultimately found to have anoxic brain damage involving his motor cortex bilaterally. He is breathing on a ventilator. Which of the following statements is most likely to be true?

    A. He will never be able to breathe on his own again because of the damage to the motor cortex.
    B. He will be able to breathe without the ventilator if he is given a phrenic nerve pacemaker.
    C. He will be able to breathe on his own because his brainstem is still intact.
    D. He will be able to breathe on his own when he is awake but not when he is asleep.

2. In the 1950s, there were several severe polio epidemics in the United States. Many people who were infected with the polio virus developed failure of the respiratory system and were kept alive in an iron lung, an airtight cylinder that encased the patient's body, except for the head, which protruded from the device. The iron lung was attached to a large vacuum that cycled the pressure inside the iron lung from atmospheric pressure to a negative pressure and then back to atmospheric pressure. This device led to the creation of a negative pressure in the pleural space on each cycle. Which part of the respiratory system was being supplemented primarily by the iron lung?

    A. the central pattern generator
    B. the peripheral nerves
    C. the airways
    D. the muscles
    E. the alveoli

3. A 65-year-old woman comes to her doctor with a complaint of increasing shortness of breath on exertion over the past 6 months. She has had multiple respiratory infections over the past 10 years and was diagnosed with a condition in which the cartilage is weakened in the lungs. Pulmonary function testing shows that there is increased airways resistance during exhalation. Which portion of the bronchial tree is likely to be affected?

    A. the conducting airways
    B. the transitional zone
    C. the gas-exchanging zone

4.   A 55-year-old man with a long smoking history is found to have a mass in the left lung. The mass is very close to the hilum, the central part of the lung where the bronchi leading to the lobes are found. There is no evidence of spread, however, and the patient is offered surgical therapy. Because of the location of the mass, the surgeon must do a pneumonectomy (removal of the entire left lung). What do you expect to happen to the pulmonary artery pressure in the remaining lung after surgery?

    A.   The pulmonary artery pressure will decrease by 50%.
    B.   The pulmonary artery pressure will stay the same.
    C.   The pulmonary artery pressure will double.
    D.   The pulmonary artery pressure will triple.

5.   A patient with severe pneumonia involving both lungs is in respiratory failure in the intensive care unit. He has an endotracheal tube in his trachea and is connected to a mechanical ventilator. In an effort to improve his oxygen and carbon dioxide levels, the doctor increases the size of each breath delivered by the ventilator. This results in an increase in the alveolar pressure to very high levels, in some cases to levels greater than the associated capillaries. How will this change in the ventilator affect the matching of ventilation and perfusion in the lung?

    A.   It will improve the matching of perfusion and ventilation.
    B.   It will have no effect on the match of perfusion and ventilation.
    C.   It will worsen the matching of perfusion and ventilation.

chapter 3

Statics: Snapshots of the Ventilatory Pump

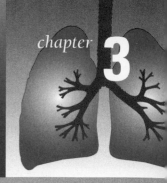

CHAPTER OUTLINE

LEARNING OBJECTIVES

- **To describe the forces that determine lung volumes.**
- **To describe the interactions between chest wall, pleura, and lungs that are responsible for the passive properties of the respiratory system.**
- **To outline the principles behind the concept of compliance and illustrate their application to the respiratory system.**
- **To describe the role of surface tension in the inflation and deflation of the lung.**
- **To define the role of surfactant in the stabilization of alveoli.**

Throughout life, the respiratory system is always "on duty" as gas is moved from the atmosphere down to the alveoli and back out again. Every minute of every day, the ventilatory pump moves 5 L of air in and out of the lungs (more during exercise and other activities). This process requires the integrated action of the chest wall, pleura, and lungs. As we begin our study of the ventilatory pump, some questions come to mind. What determines how much air we can breathe into the lungs? Why can't we exhale all the air out of our lungs? What determines the size of our lungs at the end of a normal breath? Although the ventilatory pump is constantly in motion, we begin our examination of its function by freezing it for a moment in time and examining its properties when there is no air moving into or out of the lungs. The study of the respiratory system during these "snapshots" is termed **statics**.

Lung Volumes and the Balance of Forces

DEFINING THE FORCES: CHEST WALL AND LUNGS

The chest wall, composed of the bones, muscles, and connective tissue of the thorax, has **elastic properties**. This means that when forces are applied to the chest wall and it is stressed or moved from its resting position, the chest wall resists the movement. When the stress is relieved, the chest wall returns to its resting or unstressed position (much as an elastic band resists being stretched and returns to its resting position when the force that is stretching it is removed). This resting position, also known as its equilibrium position, defines a volume because the chest wall is a three-dimensional structure. The chest wall resists deformation from this volume; if you apply a force to the chest wall to make it bigger or smaller and then release the force, the chest wall will return to its resting volume. Throughout much of the breathing that we do in our daily lives, the chest wall is at a volume that is smaller than its isolated, unstressed position (Fig. 3-1). Thus, at these volumes, it wants to spring out to resume its equilibrium configuration.

Similar to the chest wall, the lungs also have elastic properties. If you removed the lungs from the body, the lungs would like to collapse to their resting or unstressed position, a position that is equivalent to their minimum volume (see Fig. 3-1). In other words, the isolated lungs in their resting position are at their smallest possible volume; in a live human being, the lungs are always stretched above their equilibrium position and exert a collapsing force. To inflate the lung one must apply a force to overcome this **elastic recoil**

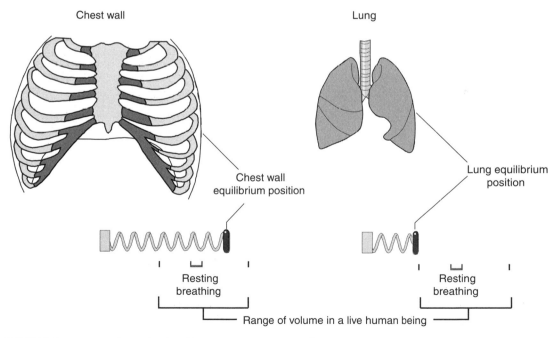

FIGURE 3-1 Resting positions of isolated lung and chest wall. The chest wall and the lungs have elastic properties. When the volume of either is increased from its resting position, it tends to resist that change in volume and, if the external force that is producing the change in volume is removed, it will return to its resting position. The chest wall also resists reductions in volume below its resting position; if the external force producing the change in volume is removed, it will return to its resting position. In contrast, in young, healthy lungs, the resting position is at a volume that represents an "airless" state (i.e., all the air has exited from the lungs); one cannot force it to a lower volume. Note that the resting position of the chest wall is near total lung capacity and the resting position of the lungs is below residual volume. **Also see Animated Figure 3-1.**

of the lung tissue (and the surface forces active in the alveoli; more about this later). When the force is removed, the lung will recoil and gas will be expelled.

Use Animated Figure 3-1 to move the springs representing the isolated lungs and isolated chest wall away from their equilibrium positions. Then release them to see how the forces bring them back. Pay special attention to the relationship between the equilibrium positions and the range of lung/chest wall volume in a live human being, including the range of resting breathing. Note that the chest wall tends to spring outward over most of the range of volume, and the lungs tend to recoil inward over the entire range of volume. Before reading on, think about how these forces would interact if we joined the chest wall and lungs together. At what volume do the lung and chest wall forces balance each other? We answer this question and address its significance shortly.

Of course, the chest wall and the lungs do not operate in isolation. The motion of one affects the other via the pleural space. As described in Chapter 2, the lungs are surrounded by a thin layer of tissue called the **visceral pleura**, and the chest wall is lined by a similar layer of tissue called the **parietal pleura**. Between these layers of tissue is the very small **pleural space**, which contains just a few milliliters of fluid. From a given unstressed position, if inspiration were initiated and the chest wall began to move outward to a larger volume, the pleural space would enlarge if the lungs did not also increase in volume. Similarly, during a passive exhalation, air exits the lungs, and the volume of the lungs decreases. If the chest wall volume did not change, the volume of the pleural space would enlarge during exhalation. The pleural space, however, is a closed space (i.e., no air or fluid moves into or out of it). The enlargement of the pleural space, therefore, results in the creation of a vacuum or negative pressure. This negative pressure is transmitted to the chest wall and lungs and intimately links the motion of one to the other. Consequently, the size of the pleural space never changes significantly as long as it remains a closed system. During normal breathing in a person at rest, the pleural pressure typically ranges during the respiratory cycle between -8 cm H_2O at end inspiration and -3 cm H_2O at end expiration.

?   **THOUGHT QUESTION 3-1: What happens when the pleural space is disrupted and the relationship between the chest wall and lung is broken?**

PRESSURE–VOLUME CHARACTERISTICS OF THE CHEST WALL, LUNGS, AND RESPIRATORY SYSTEM

Because the chest wall and lungs have elastic properties and resist changes in volume from their resting positions, a force must be applied to them to produce a change in volume. These forces can be conceptualized as unequal pressures on either side of the walls of the chest cavity or the lung. We call the pressure across the wall of a structure the **transmural pressure**. Mathematically, transmural pressure is expressed as follows:

$$P_{TM} = P_{inside} - P_{outside}$$

where P_{TM} is the transmural pressure, P_{inside} is the pressure on the inside of an enclosed structure, and $P_{outside}$ is the pressure on the outside of an enclosed structure. By convention, the transmural pressure is always expressed as inside minus outside. Positive transmural pressures, therefore, are associated with forces that tend to expand or increase the volume of a structure. Alternatively, negative transmural pressures are associated with collapsing forces that tend to decrease the size or volume of a structure. The resting or unstressed position of the structure is the volume at which the transmural pressure is zero (Fig. 3-2).

A number of elements of Figure 3-2 must be examined closely. First, look at the axes. The x axis is the transmural pressure. As noted, a positive transmural pressure indicates

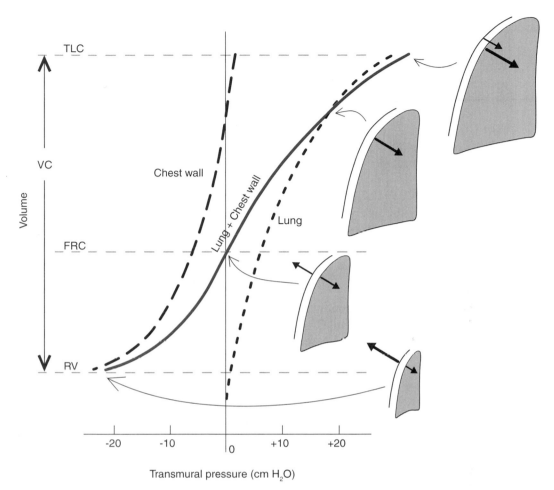

FIGURE 3-2 Pressure–volume characteristics of the respiratory system. This diagram shows the passive pressure–volume characteristics of the lung and chest wall in isolation, as well as the two integrated together as the respiratory system. Note that the resting position of the lungs is at a volume below the residual volume (RV), and the resting position of the chest wall is close to the total lung capacity (TLC). The schematics of the lung and chest wall show the forces exerted by the lungs and chest wall at the different volumes. At functional residual capacity (FRC), the forces exerted by the chest wall and the lungs are equal in intensity and opposite in direction. Thus, this is the resting position of the respiratory system. At volumes above FRC, the balance of forces is such that the respiratory system wants to get smaller; at volumes below FRC, the balance of forces pushes the system toward a higher volume, that is, back to FRC. VC = vital capacity. **Also see Animated Figure 3-2.**

that the pressure inside the structure is greater than the pressure outside the structure (a positive transmural pressure is a distending pressure). A negative transmural pressure indicates that the pressure outside the structure is greater than inside (this leads to compression of the structure). The point at which the curve crosses the 0 pressure line is the resting or unstressed volume of the structure.

Now look at the y axis, the volume axis. The volume axis is expressed in terms of the percent of **total lung capacity (TLC)**, which is defined as the volume of air in the lungs at the end of a full inspiration. **Residual volume (RV)** is the volume remaining in the lung at the end of a forced exhalation. The difference between TLC and RV is the **vital capacity (VC)**. Mathematically, the TLC is the sum of the RV and VC.

$$TLC = RV + VC$$

The resting position of the chest wall in isolation is at a volume that is approximately 75% to 80% of the TLC. Volumes below this level require a force to be applied to the chest wall to compress it (negative transmural pressure); volumes above this level require a force to be applied to expand the chest wall (positive transmural pressure). At the resting or unstressed position, the volume of the lung is near 0 (i.e., virtually no air in the alveoli). To expand the lung above this volume, a positive transmural force must be applied.

The line representing the respiratory system (lung + chest wall) in Figure 3-2 is the algebraic sum of the chest wall and lung pressure–volume curves. Note that the point at which the respiratory system crosses the 0 pressure line denotes the volume at which the force exerted by the tendency of the lung to recoil inward is equal to and opposite the force exerted by the chest wall to spring outward. This volume is referred to as **functional residual capacity (FRC)** and represents the volume of the lungs at the end of a normal, relaxed exhalation. At this volume, the inward or deflating force of the lungs is *balanced* by the outward or inflating force of the chest wall (recall that this balance of forces is mediated via the pleural space).

Use Animated Figure 3-2 to view the changes in the lung and chest wall forces (shown as force vectors directed inward or outward) over the range of lung volume. Pay special attention to the forces at FRC and to how the net force (i.e., sum of lung and chest wall forces) varies above or below this volume. You may find it helpful to reread the previous section while using the Animated Figure to view the associated forces at each volume. You can watch a detailed explanation as part of the Animated Figure by clicking on the ShowMe button.

THE FORCES: ADDITIONAL FACTORS TO CONSIDER

Elastic forces, although they explain much of the static properties of the lungs, do not provide all the information you need to understand lung volumes and the interactions of the chest wall and the lungs.

Muscles

We defined the resting position of the lungs, chest wall, and respiratory system as the volume that the structure would achieve when the transmural pressure is 0. What moves the system from the resting volume? How is the force generated to change the transmural pressure from 0?

Inspiration is an *active* phenomenon that requires the use of the inspiratory muscles (remember that the net force vector of the lungs plus the chest wall is directed inward above FRC). The tension generated in a muscle when stimulated by a neurological impulse depends on the length of the muscle at the moment at which the neurological stimulus reaches the muscle. The relationship between the length of a muscle and the tension generated in the muscle can be determined with in vitro experiments in which isolated muscle strips are examined (Fig. 3-3).

Figure 3-3 represents an isometric contraction in which the muscle generates force but is not allowed to shorten (analogous to pushing against an unmoving brick wall—your muscles generate tension despite staying the same length). A muscle can generate greater tension (force) when stimulated from a higher initial length, a relationship shown by the positive slope of the isometric contraction tension curve. The passive component is due to the elastic elements of the muscle being stretched. Based on this length–tension relationship, where do you think the diaphragm is in the most advantageous position for force generation—at end inspiration or at end expiration? Although this question may be difficult to answer if you do not yet have some familiarity with the anatomy of the diaphragm, think about it before reading on and finding out the answer.

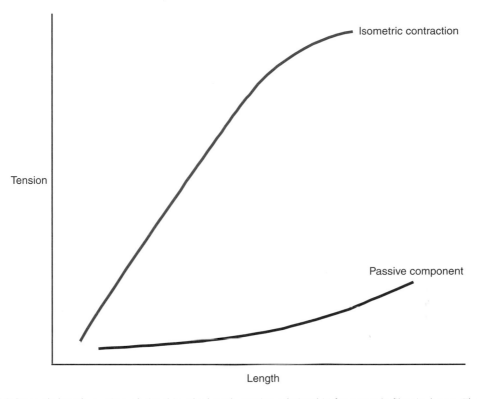

FIGURE 3-3 Muscle length–tension relationship. The length–tension relationship for a muscle fiber is shown. The magnitude of the tension generated is related to the length of the muscle at the time that it is stimulated to contract. Up to a point, the longer the muscle fiber, the greater the tension generated. Ventilatory muscles demonstrate this length–tension relationship.

The diaphragm, the primary inspiratory muscle, is dome shaped at the end of expiration; the peak of the diaphragm comes to a point at a level near the xiphoid process. Laterally, the diaphragm is in close proximity to the rib cage, an area called the *zone of apposition* because the diaphragm and chest wall are literally apposed or up against one another here (Fig. 3-4). As the diaphragm contracts, it shortens and moves downward or caudally, leading to an increase in intra-abdominal pressure and outward movement of the rib cage (see Fig. 3-4). Use Animated Figure 3-4 to view the movement of the rib cage and the change in length of the diaphragm over the range of breathing. At low thoracic volumes, for example, at the end of exhalation, the diaphragm along with the external intercostal muscles, the other major inspiratory muscles, are relatively long. Thus, the tension generated in the muscles is high when stimulated by a neurological impulse from the controller at FRC. This results in more effective pressure generation by the ventilatory pump, thereby creating appropriate inspiratory flow (by moving the chest wall outward, pleural pressure becomes more negative; this negative pressure is transmitted to the alveoli and produces flow into the lungs; more on this in Animated Fig. 3-9). In contrast, at high lung volumes, at the end of inspiration, the inspiratory muscles are shortened and are less effective at generating tension for a given neural stimulus. Consequently, inspiratory flow at high lung volumes is less than at low lung volumes for the same neurological stimulus to the muscles (more on this in Chapter 4).

The length–tension relationship shown in Figure 3-3 is a general property of skeletal muscles and therefore applies to the ventilatory muscles, as well as being characteristic of myocardial muscle cells (the length–tension relationship for the heart is an application of

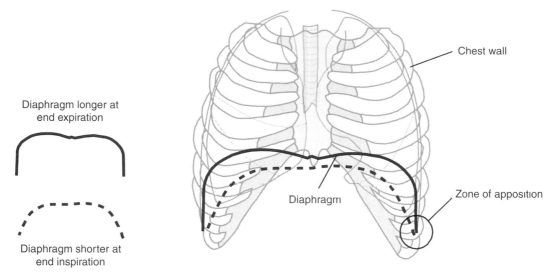

FIGURE 3-4 Diaphragm position and function. The position of the diaphragm is shown at two points in the respiratory cycle, end expiration and end inspiration. Note that the diaphragm is higher in the thorax near the end of expiration. The diaphragm is relatively "long" in this position. Thus, when stimulated at the beginning of the next breath, the diaphragm will be able to generate a high tension. In contrast, near the end of inspiration, the diaphragm is shortened and is less effective at generating tension. **Also see Animated Figure 3-4.**

this principle to a three-dimensional structure, the ventricle, and is termed the *Starling curve*; if you have not yet studied cardiovascular physiology, remember this concept for Chapter 9, which provides some of the basics of cardiac function during exercise).

Changes in lung volumes occur in physiologic and pathologic states in ways that alter the length and, therefore, the performance of the inspiratory muscles. For example, under conditions in which the individual must increase the amount of air moving into and out of the lungs, such as during exercise, the system responds partly by exhaling more fully than normal. This has the effect of reducing the lung volume at the end of exhalation, thereby lengthening the diaphragm and enhancing its ability to contract just before the next inhalation. In contrast, some disease states, such as emphysema, are associated with a larger than normal lung volume at the end of exhalation, a finding that places the diaphragm in a shortened position and impairs its ability to generate tension during inspiration.

Unlike inspiration, expiration is usually a passive phenomenon during relaxed breathing. When the inspiratory muscles stop contracting at the end of inspiration, the normal elastic properties of the lung and, at high thoracic volumes, the recoil of the chest wall, lead to flow of gas out of the lung. During periods of increased ventilation—for example, during exercise or when a person needs to generate high expiratory flow, as during a cough—expiration is the consequence of a combination of the passive recoil properties of the lung and chest wall as well as the active contraction of expiratory muscles. The primary expiratory muscles are the muscles of the abdominal wall and the internal intercostals. They are at their longest position and, hence, best able to generate tension, when the thoracic volume is high, as occurs after a full inhalation. Expiratory muscles are in their shortest position and are least able to generate tension at the end of exhalation.

Airway Closure

In children and young adults, RV is determined solely by the balance of forces exerted by the chest wall (outward), the lungs (inward), and the expiratory muscles (inward). In older individuals (often beginning after age 40), however, the elasticity of the lungs begins to diminish. The diameter of small airways in the lung is determined partly by the volume of the lung; the higher the lung volume, the bigger the diameter of the airways. Imagine a mesh stocking with a complex latticework of fibers surrounding holes or openings in the material. As you stretch the stocking by pulling on the ends, the size of each of the openings in the middle increases. When the lung is stretched during inhalation, the diameter of the airways increases. During exhalation, the stretch is reduced and the airways decrease in size. The residual "pulling" by the surrounding lung tissue, however, keeps the airway from collapsing completely as transmural pressure becomes negative. As elastic recoil of the lung diminishes with age, the small airways are more susceptible to collapse during exhalation.

During exhalation, the lung becomes progressively smaller and, if the person tries to force as much gas as possible out of the lungs (i.e., to exhale to RV), energy is needed to drive the chest wall below its resting position. Depending on the stiffness of the chest wall (which increases with age) and the elasticity of the lung (which decreases with age), pleural pressure may become positive as expiratory muscles are activated or **recruited**. As the lungs become smaller, the transmural pressure in the small airways becomes negative, predisposing some of these airways to collapse. When the airway collapses, gas distal to the point of collapse—in other words, the gas out toward the alveoli—cannot exit. The gas is described as *trapped* behind the collapsing or narrowed airway. The more air that is trapped in this manner, the greater the RV. In older adults, therefore, as well as in people with disease states such as emphysema that reduce the elastic recoil of the lungs, the balance of forces does not completely determine the RV. Air trapping must also be considered.

Compliance

As one tries to inflate or distend an object such as a balloon or a ball, one has to increase the pressure inside the object relative to the pressure outside it, that is, one has to create a positive transmural pressure. To go from one volume to a second, larger volume, one has to further increase the transmural pressure. The relationship between the change in volume and the change in the pressure needed to achieve that change in volume is called the **compliance** of the object.

$$\text{Compliance} = \frac{\Delta \text{ Volume}}{\Delta \text{ Pressure}}$$

$$C = \frac{\Delta V}{\Delta P}$$

An object with a low compliance is relatively stiff; it requires a large change in pressure to achieve a given change in volume. It is important to remember that the pressure that leads to a given volume is the pressure across the wall, the transmural pressure, of the object, and not the pressure within the object, to which one may refer as the **intracavitary pressure** (Fig. 3-5).

Use Animated Figure 3-5 to vary the transmural pressure for a floppy balloon, a normal balloon, and a stiff balloon and observe the resulting changes in volume. For a given change in transmural pressure, which of these balloons do you think undergoes the greatest change in volume? That balloon (the thin, floppy one) has the highest compliance of the three.

THOUGHT QUESTION 3-2: What happens to the pressure and volume inside an inflated balloon when it is placed under water? Why?

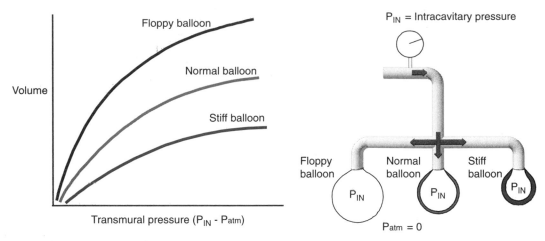

FIGURE 3-5 Compliance. This diagram shows three balloons, all connected by tubes so that the pressure within each balloon, the intracavitary pressure (P_{IN}), is the same. The pressure outside the balloons is atmospheric pressure (P_{atm}) which, by convention, is expressed as 0 cm of water or millimeters of mercury. The transmural pressure, the pressure inside minus the pressure outside the balloon ($P_{IN} - P_{atm}$) is, therefore, the same for each balloon as well. Despite the same transmural pressure, the balloons contain different volumes of gas. The volume of gas is proportional to the **compliance** of the balloon, a measure of its "stiffness." The graph shows the relationship between volume and transmural pressure for each balloon. The less compliant or stiffer the balloon, the lower is the slope of the curve. The slope of the curve ($\Delta V/\Delta P$) is equal to the compliance. **Also see Animated Figure 3-5.**

Because the pressure outside the object may be 0 (by convention, atmospheric pressure is taken as 0), the transmural pressure and the intracavitary pressure may be the same. In the context of the respiratory system, the pressure "outside" the lung is the pleural pressure which, under normal conditions, is approximately −3 to −5 cm H_2O at FRC. In individuals with obesity or those with respiratory failure who are being sustained on mechanical ventilators, however, the pleural pressure may be positive and quite different from 0. In these circumstances, it is critical to remember that changes in lung volume relate to changes in transmural pressure. Failure to determine pleural pressure as part of the calculation of transmural pressure in these circumstances can lead to errors in assessment of the patient's respiratory status. (Note: the concept of transmural pressure as a distending pressure is also critical in cardiovascular physiology when one assesses cardiac function in patients in whom pleural pressure is not close to 0.)

When one inflates the lungs, the change in transmural pressure needed to attain a given change in volume depends on the compliance of the respiratory system. Many disease states can affect the compliance of the lungs. Emphysema, for example, destroys lung tissue and reduces elastic recoil of the lung, leading to an increase in compliance. In contrast, some inflammatory conditions lead to scarring or fibrosis with deposition of stiff connective tissue in the lung and a decreased compliance (Fig. 3-6).

THOUGHT QUESTION 3-3: What are the differences in compliance as one inflates the lung between RV and TLC in the three conditions shown in Figure 3-6? Calculate the average compliance for each condition.

Diseases of the chest wall may also alter compliance. Curvature of the spine, or scoliosis, for example, as well as obesity (remember, the abdomen is the inferior border of the chest wall), are often associated with a decrease in chest wall compliance.

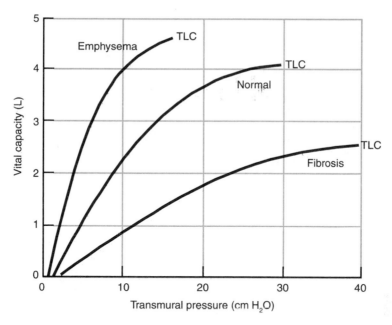

FIGURE 3-6 Lung compliance. The pressure–volume characteristics for three lungs are shown. One lung is from a patient with emphysema, one is from a normal individual, and one is from a patient with pulmonary fibrosis. Emphysema is a process that results in reduced elastic recoil of the lung, which leads to an increase in compliance. Pulmonary fibrosis is associated with scarring in the lung, a process that leads to decreased compliance. Note that the slope of the curve ($\Delta V/\Delta P$) is equal to the compliance. TLC = total lung capacity. (Adapted with permission from Murray JF: *The Normal Lung*, 2nd ed. Philadelphia, WB Saunders, 1986, p. 87.)

LUNG VOLUMES: A CLOSER LOOK

In Figure 3-2 and Animated Figure 3-2 we examined the pressure–volume characteristics of the lungs, chest wall, and respiratory system. Consider now the forces that are balanced to determine TLC, FRC, and RV. Let us start at FRC and ask an individual to breathe into a water **spirometer**, a device that collects expired gas under a drum that is partially suspended under water and transforms the motion of the drum as the patient inhales and exhales into a graphical representation of the change in lung volume as a function of time (Fig. 3-7 and Animated Fig. 3-7). You instruct the person, at the end of a relaxed exhalation, to take a deep breath in until she cannot get any more air into her lungs. The volume of the respiratory system at this point is the TLC and is determined by the balance of the forces exerted by the chest wall and lungs inward and the force of the inspiratory muscles to expand the system. The amount of air inhaled from FRC to TLC is termed the **inspiratory capacity** (not labeled in the figure).

Next you instruct the patient to exhale until she cannot get any more air out of her lungs. The volume of the respiratory system at this point is the RV and is determined by the balance of the force exerted by the chest wall to spring out and the force of the lungs and the expiratory muscles to make the system smaller (remember that collapse of small airways and the resulting air trapping may also play a small role in determining RV in the normal adult). The amount of air exhaled from TLC to RV is termed the vital capacity (VC). The difference in volume between FRC and RV is termed the **expiratory reserve volume (ERV)** (not labeled in the figure). The **tidal volume** (usually abbreviated as V_T or just VT) is the amount of air inhaled during a normal breath.

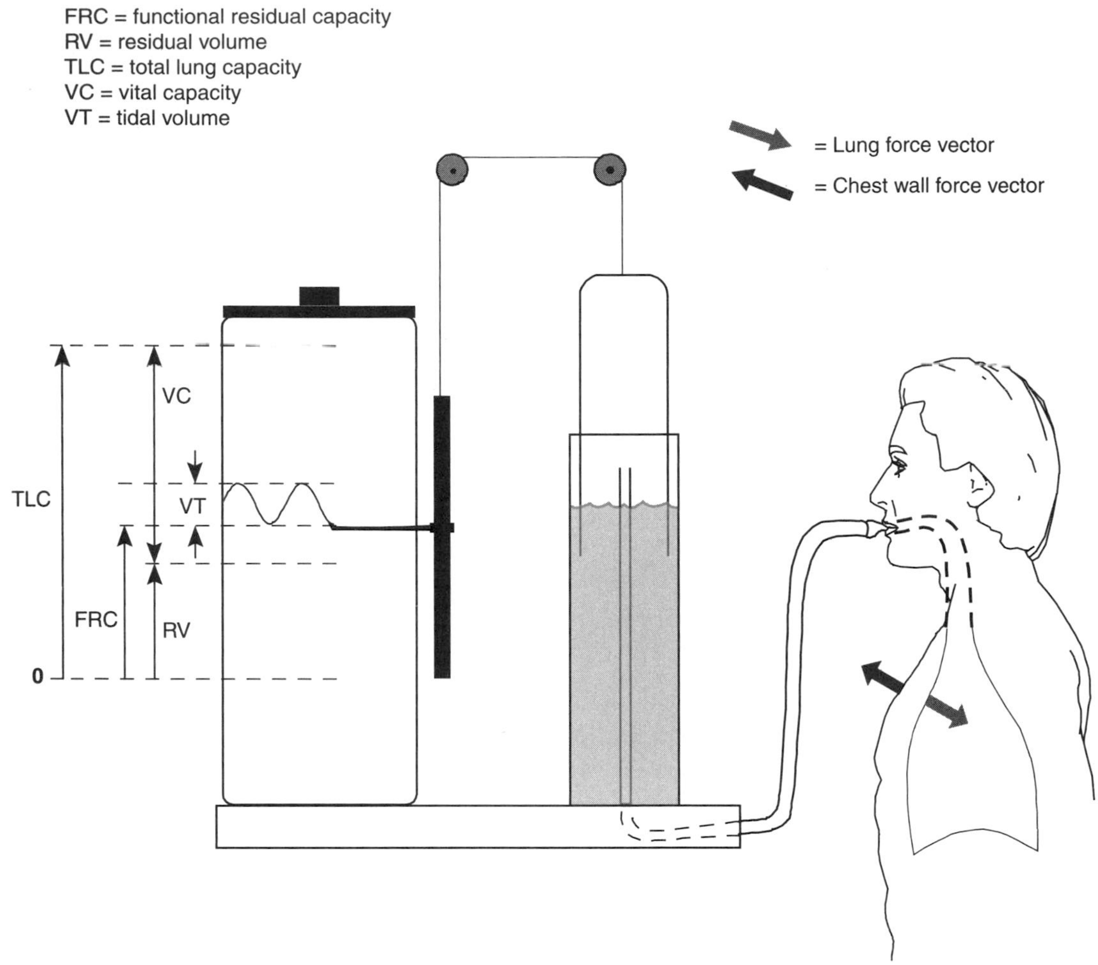

FRC = functional residual capacity
RV = residual volume
TLC = total lung capacity
VC = vital capacity
VT = tidal volume

FIGURE 3-7 Spirometry and lung volumes. A patient is shown breathing on a water spirometer. As the patient exhales, the expired gas forces up the spirometer drum; during inhalation, the drum descends. The movement of the drum is transformed into movement of a pen, and the system is calibrated so that the vertical motion can be translated into changes in volume. The figure depicts the patient at functional residual capacity. Note that the force exerted by the lungs inward (the elastic recoil of the lungs) is equal in magnitude but opposite in direction to the outward recoil force exerted by the chest wall. **Also see Animated Figure 3-7.**

Use Animated Figure 3-7 along with Table 3-1 to understand the divisions of lung volume. Animated Figure 3-7 initially shows a person performing normal relaxed breathing into a water spirometer. The height of the waveform tracing represents tidal volume. You can use the Show VC button to instruct the subject to breathe in as fully as possible and then breathe out as fully as possible (the sequence discussed previously). The height of the resulting waveform tracing represents the vital capacity. Also note how the lung and chest wall force vectors vary over the breathing cycle. Based on your prior knowledge, what is the relationship between the forces exerted by the lungs and chest wall at FRC? If this question seems difficult, you may find it helpful to revisit Animated Figure 3-2.

When air is exhaled with as much force as the person can generate from TLC to RV, we term the volume the **forced vital capacity**, or **FVC**. During an FVC maneuver, we can also measure the amount of gas that is exhaled in the first second of the FVC. This volume is termed the **forced expiratory volume in one second (FEV$_1$)**. Individuals with obstructive lung diseases such as asthma, emphysema, and chronic bronchitis typically have abnormally low FEV$_1$ because of high airway resistance and, in the case of emphysema, a reduced lung elastic recoil that results in abnormally low force generation during exhalation. The ratio of the FEV$_1$ to the FVC, which should be greater than 0.7 in most healthy people, is one of the measures used to diagnose airways disease.

THOUGHT QUESTION 3-4: A patient has a reduced inspiratory capacity and TLC. FRC, ERV, and RV remain unchanged. Is the problem caused by a disorder of the chest wall, the lungs, or the diaphragm? Why?

Knowledge of the balance of forces that determine lung volumes is critical to your understanding of the patterns of abnormality seen with different disease states.

Measurement of Lung Volume

With the spirometer shown in Figure 3-7 and Animated Figure 3-7, we can calculate changes in volume from FRC to TLC (i.e., the inspiratory capacity) and from TLC to RV (i.e., the vital capacity). This allows us to quantify the vital capacity, inspiratory capacity, and expiratory reserve volume (Table 3-1). To know the true values for FRC, TLC, and RV, however, we need to know the absolute volume for at least one of them. With that information in hand, we can then derive the other two by looking at changes in volume with a spirometer. For example, if you know the value for TLC, you could ask the person to exhale from TLC until she could not exhale any more (perform a vital capacity maneuver). The TLC minus the VC gives you the RV.

Given the choice between measuring TLC, FRC, or RV directly, which would you prefer to measure? Which would likely give the most reproducible results?

Functional residual capacity is typically measured directly. Its measurement is more consistent than the measurement of TLC and RV because it is not dependent on the use of ventilatory muscles. FRC is the volume that is achieved at the end of a relaxed

TABLE 3-1 Lung Volumes		
LUNG VOLUME	**DEFINITION**	**MATHEMATICAL EXPRESSION**
Total lung capacity	Volume at the end of a maximal inhalation	TLC = RV + VC
Functional residual capacity	Volume at the end of a normal exhalation	
Residual volume	Volume at the end of a maximal exhalation	RV = TLC − VC
Inspiratory capacity	Volume between FRC and TLC	IC = TLC − FRC
Expiratory reserve volume	Volume between FRC and RV	ERV = FRC − RV
Vital capacity	Volume of gas exhaled from TLC to RV	VC = TLC − RV

ERV = expiratory reserve volume; IC = inspiratory capacity; FRC = functional residual capacity; RV = residual volume; TLC = total lung capacity; VC = vital capacity.

exhalation when the force exerted by the elastic recoil of the lung is equal to and opposite the force exerted by the chest wall to spring outward (recall Animated Figs. 3-2 and 3-7). To reach TLC requires the patient to make a maximal inspiration; to reach RV requires the patient to make a maximal exhalation. In either case, you must depend on the cooperation of the person doing the test to achieve accurate measurements. If the individual were to give less than a maximal effort, all other volumes derived with the spirometer would be in error.

To measure the volume of gas in the lung at FRC, one needs a marker gas that is inert and relatively insoluble in blood. This permits the gas to stay in the alveoli because it will not be absorbed into the blood or metabolized as it is inhaled by the patient. Thus, the gas is diluted as it moves into the lung and mixes with the air already in the lung. The technique commonly used to measure FRC is termed **helium dilution** (Fig. 3-8).

The person breathes in a relaxed manner on a mouthpiece connected to a valve that leads to a container with a known volume and a known concentration of helium. After the individual becomes comfortable on the apparatus and the technician notes the person is at FRC, the valve is turned so that the individual is now breathing from and into the container. The concentration of helium in the container is monitored as it gradually decreases from its initial concentration. After 2 to 3 minutes in an individual with normal

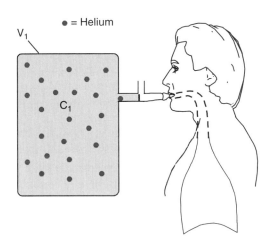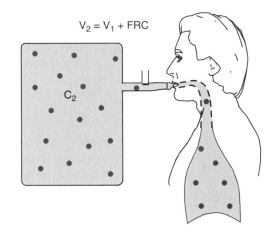

A Before opening valve

B Valve opened at FRC.
Now shown after several minutes.

$$C_1 \times V_1 = C_2 \times V_2 = C_2 \times (V_1 + FRC)$$

FIGURE 3-8 Helium dilution measurement of functional residual capacity (FRC). **A,** The subject is breathing on a mouthpiece connected to a valve. The valve is connected to tubing that links to a container with a known volume (V_1) and concentration (C_1) of helium. When the patient is at FRC, the valve is turned, and the subject breathes in and out from the container (shown in **B**). **B,** Graphic representation of the concentration of helium in the system during the helium dilution test. The container of known volume (V_1) contains a given concentration of helium (C_1). When the subject is at FRC, the valve is turned, and the subject begins to breathe in and out from the container. The helium in the box is now inhaled into the lungs, and the concentration of helium in the container begins to decrease. After several minutes, the helium is distributed evenly between the container and the lungs, and a new steady-state concentration of helium is achieved (C_2). With values known for C_1, V_1, and C_2, the value of FRC can be calculated. Note that while the subject is breathing into the container, oxygen is added to the container at a rate equal to the oxygen being consumed each minute by the subject. In addition, the carbon dioxide exhaled is removed from the system so that it does not accumulate in the container.

lungs, the concentration of helium levels off as it comes into equilibrium with the gas originally in the lungs. You know the original volume (V_1) and concentration of helium (C_1), and you now know the final concentration of helium (C_2). With this information, you can calculate the new volume (container + lungs, or V_2) into which the helium is distributed:

$$C_1V_1 = C_2V_2$$

Subtraction of the volume of the box (V_1) from the new volume of distribution (V_2) gives you FRC:

$$V_2 - V_1 = FRC$$

? THOUGHT QUESTION 3-5: What are the limitations of the helium dilution technique for the measurement of FRC in a patient with areas of lung that communicate poorly with the central airways?

There are two alternative methods for measuring FRC. The first is termed the **nitrogen washout test**. This test makes use of the fact that approximately 79% of the gas we breathe from the atmosphere is nitrogen (the remaining 21% is primarily oxygen). Because there is no net movement of nitrogen from the lungs into the blood, the amount of nitrogen exhaled is the same as the amount inhaled. To do this test, ask a person to breathe 100% oxygen, starting at FRC (V_1 in this example). The nitrogen in the lungs will be gradually replaced by oxygen as the person exhales nitrogen and replaces it with oxygen. Measure the volume of exhaled gas for a fixed period of time (V_2) (e.g., 5 minutes) after oxygen breathing is started, along with the concentration of exhaled nitrogen in that gas (C_2). At the end of the 5-minute collection of gas, measure the nitrogen concentration of the very last gas exhaled (C_3). This represents the concentration of the nitrogen still remaining in the alveoli. The original volume of the lung at the time that oxygen breathing was started can be calculated as follows:

$$V_1 \times 80 = (V_1 \times C_3) + (V_2 \times C_2)$$

The nitrogen washout method has the same limitation as does the helium dilution technique. You are measuring only the lung units that have good communication with the central airways.

To deal with this issue of lung units that do not communicate well with the central airways, a third method called **body plethysmography** can be used (Fig. 3-9). The body plethysmograph is a device that makes use of Boyle's law to measure the volume of gas within the thorax. In this test, a person is seated within an airtight box and breathes on a mouthpiece connected to a tube that links the individual to equipment outside the box. The volume of the box is known, and the pressure within the box is measured. When the person is at FRC, a valve is closed in the tube on which the person is breathing and he is instructed to make quick inspiratory and expiratory efforts against the closed airway. During this time, the pressure changes in the person's airways (and hence, the lungs because there is no flow) are measured as he compresses and decompresses the gas in the thorax. Simultaneously, the changes in the pressure of the box, which occur as the person's chest expands and contracts during the breathing efforts, is also measured. Knowing the volume of the box, the changes in pressure in the box and in the individual's airway, the volume of gas in the thorax can be calculated.

Boyle's Law for box: $P_1V_1 = P_2(V_1 - \Delta V) \rightarrow$ by measuring P_1, P_2, knowing V_1 (volume of box) $\rightarrow$ can get ΔV

Boyle's Law for lungs: P_3 (FRC) $= P_4$ (FRC $+ \Delta V) \rightarrow$ by measuring P_3, P_4, using calculated $\Delta V \rightarrow$ can get FRC

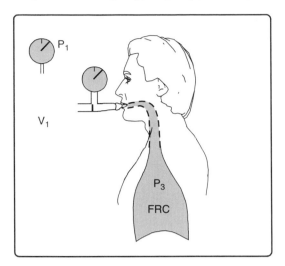

Airtight box, subject inside
Close mouthpiece valve at FRC

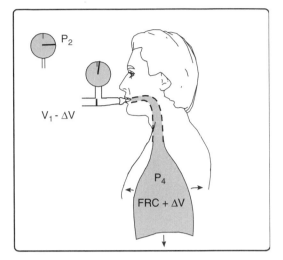

Subject tries to inhale against closed valve,
changing lung by ΔV and box by $- \Delta V$ volume and
lowering/raising pressure in lung/box

FIGURE 3-9 Body plethysmography. In this technique, the subject is placed in an airtight box and breathes through a mouthpiece with a pressure gauge. When the subject is at the end of a normal expiration (functional residual capacity [FRC]), the mouthpiece valve is closed and the subject is instructed to make attempts to inhale and exhale. Boyle's law, which states that pressure times volume is constant for a confined gas (at a constant temperature), can be used as shown to calculate FRC. The figure shows the attempt at inhalation and the method for calculating FRC.

Alveoli: Distending Pressures and the Stability of Lung Units

To this point, we have been examining the lung at a "macro" level. We have reviewed the properties of the chest wall and lungs as well as the interactions between these structures that help us to understand the balance of forces that guides inhalation and exhalation. These forces determine the lung volumes FRC, TLC, and RV. A more complete understanding of these forces, however, depends on an examination of the lung at a "micro" level; we must look at the properties of the alveolus and how it behaves during the respiratory cycle.

TRANSMURAL AND TRANSPULMONARY PRESSURE

Earlier in the chapter, we introduced the concept of transmural pressure. We will now extend that discussion in the context of the alveolus. *Transmural pressure* is a general term for pressure across the wall of an object (transmural means "across the wall") and can be described by the following equation:

$$P_{TM} = P_{inside} - P_{outside}$$

A flexible container or object expands if there is a positive transmural pressure (pressure greater inside than outside the object) and gets smaller with a negative transmural pressure. For this reason, a positive transmural pressure is sometimes referred to as a "distending" pressure. (Remember, one can describe the transmural pressure of the alveolus, the airway, or for that matter, the right ventricle of the heart. The term is general and can apply to any structure with a wall.)

For the alveolus, the pressure inside is the alveolar pressure, P_{alv}, and the pressure outside is the pleural pressure, P_{pl}. Therefore, transmural pressure for the alveolus can be written as follows:

$$P_{TM} = P_{alv} - P_{pl}$$

The pressure in the alveolus is a result of two forces: the elastic recoil of the lung (a collapsing force, P_{el}), and the forces surrounding the lung (the pleural pressure). Thus, alveolar pressure is always greater than pleural pressure.

$$P_{alv} = P_{el} + P_{pl}$$

Substituting this expression for P_{alv} into the previous equation, the transmural pressure of the alveolus can be expressed in terms of elastic recoil pressure:

$$P_{TM} = (P_{el} + P_{pl}) - P_{pl}$$
$$P_{TM} = P_{el}$$

Thus, under static conditions, the distending pressure of the lungs—the pressure that determines the volume of the lung at that moment in time—is the same as the elastic recoil pressure.

(Clinically, you will hear the term *transpulmonary pressure* used. This is the difference between the pressure at the mouth [taken to be the 0 reference point] and the pleural pressure. *Transpulmonary pressure* is used instead of *alveolar transmural pressure* because there is no practical way to measure the alveolar pressure.)

At FRC, when no ventilatory muscles are activated, the pleural pressure reflects the interaction between the tendency of the lungs to collapse (elastic recoil) and the tendency of the chest wall, at this volume, to spring out or expand. This creates a small negative pressure in the pleura, approximately −3 to −5 cm H_2O. Because the alveolar pressure at FRC is 0 (this is the transition between expiration and inspiration, so there is no flow), the transmural pressure and the elastic recoil pressure must be between 3 and 5 cm H_2O (equal and opposite to the pleural pressure) at this volume (Fig. 3-10A). A greater transmural pressure equates with a greater lung volume and greater elastic recoil.

Let's begin to set the system in motion. The inspiratory muscles are activated, pulling the chest wall outward, and pleural pressure becomes more negative. This negative pressure is transmitted to the alveolus. Alveolar pressure, which had been 0 an instant before, is now negative as well ($P_{alv} = P_{el} + P_{pl}$). There is now a pressure differential between the airway opening and the alveolus, and air flows into the lung (Fig. 3-10B shows a snapshot of an intermediate point in the inspiration).

Use Animated Figure 3-10 to view the sequence delineated above and pay special attention to how the alveolar and pleural pressures change during the inspiration. Transmural pressure across the alveolus is indicated in the diagram. You can use the slider to move slowly back and forth through the animation and observe the pressure changes. Note how the alveolar pressure becomes negative, causing airflow, and then returns to 0 as we reach static conditions again at the end of inspiration. You can watch a detailed explanation as part of the Animated Figure by clicking on the ShowMe button.

ALVEOLAR STABILITY

The alveolus is the primary site of gas exchange in the lungs. For gas exchange to occur, the alveolus must remain open. The physiology of alveolar stability is one of the most important and difficult areas of respiratory physiology. An understanding of this topic is essential for the effective management of patients with respiratory distress syndrome, a condition that leads to alveolar collapse and severe problems related to the low blood oxygen levels that result when lung units no longer contain air.

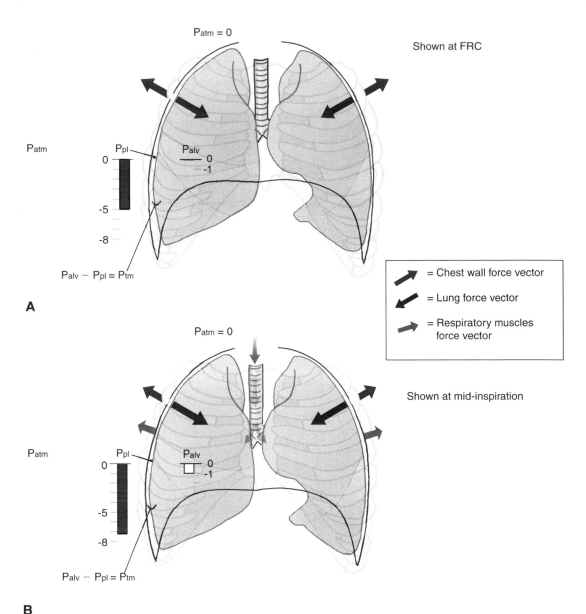

FIGURE 3-10 Setting the system in motion. **A,** The lungs and chest wall at functional residual capacity (FRC). Note that the force exerted by the lungs inward (the elastic recoil of the lungs) is equal in magnitude but opposite in direction to the outward recoil force exerted by the chest wall. Under static conditions, the alveolar pressure is equal to pressure at the airway opening (mouth) or atmospheric pressure (P_{atm}), which is 0. Pleural pressure at FRC is shown as -5 cm H_2O. **B,** We now set the system in motion by activating the inspiratory muscles (note the new force vector) and moving air into the lungs. The chest wall has moved outward because of the action of the inspiratory muscles, and pleural pressure has decreased to approximately -7 cm H_2O. Pressure in the alveolus has decreased to -1 cm H_2O, which establishes a pressure gradient between the airway opening and the alveolus, leading to flow of gas into the lungs. Alveolar pressure has not changed as much as pleural pressure because of the effect of other factors such as airway resistance, tissue resistance, and the acceleration of gas as flow is initiated (see Chapter 4). The lung is now at a higher volume, and the elastic recoil forces are greater than at FRC. The chest wall has moved closer to its resting position, so the outward recoil forces of the chest wall have diminished. **Also see Animated Figure 3-10.**

The Alveolus as a Bubble

As we begin our exploration of the forces that influence the stability of the alveolus, let us consider a soap bubble. The bubble consists of a spherical film of liquid soap surrounding gas; thus, there is a gas–liquid, or air–liquid, interface. This interface or surface gives rise to **surface tension**, which is defined as the force with which a surface contracts per unit length of surface and has the units (dynes/cm^2). A molecule within a liquid is surrounded on all sides by other similar molecules, which exert force on it. Because of the symmetry of the arrangement of the molecules, there is no net molecular movement. In contrast, at the surface of the liquid, the molecules interact with other liquid molecules in some directions (e.g., laterally in the surface of the liquid and with molecules in the subsurface layers) but not in other directions (toward the air interface). The intermolecular forces across the surface, between the liquid and gas molecules, are much weaker than the forces between molecules within the liquid. This imbalance of intermolecular forces at the surface, distinct from the uniform field that would be applied to a given molecule located in a subsurface layer, causes the molecules at the surface to be drawn toward the interior of the liquid. The net effect of this arrangement is that the surface tends to contract; the strength of that contraction is the surface tension (Fig. 3-11)

These forces allow a leaf to float on the surface of a quiet pond. Surface tension causes the surface to assume the smallest area possible, that is, the molecules at the surface come as close together as possible as they are drawn toward the subsurface molecules. In the case of a sphere such as soap bubble, this force is translated to a three-dimensional object. The tendency for the surface to assume the smallest possible area—in this case, the smallest area of a sphere—produces a force that, if unopposed, leads to the collapse of the sphere.

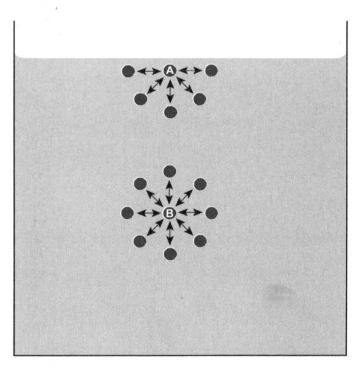

FIGURE 3-11 Surface tension. An air–liquid interface is shown. Intermolecular forces affect molecules within the liquid (molecule **B**) in all directions except at the surface (molecule **A**), where no force is felt from the direction of the air–liquid interface. This distribution of forces leads to surface tension. The liquid attempts to minimize the area of the surface as a consequence of these forces.

Law of LaPlace

If surface tension causes the sphere to assume the smallest radius possible, what would stop the soap bubble from collapsing? To answer this question, we must consider the **Law of LaPlace**, which can be stated as follows:

$$P = 2T/r$$

where P as the pressure inside the sphere, T is the tension in the wall of the sphere, and r is the radius of the sphere. What are the implications of this relationship? As the radius of the bubble decreases, it requires an increasing pressure inside the bubble to offset the tension in the wall and thereby prevent the bubble from collapsing. Alternatively, if one could reduce the tension in the wall of the bubble, one could tolerate very small radii without an increase in pressure and still avoid a collapse of the sphere.

Alveoli may be thought of as microscopic, spherically shaped objects lined with a thin film of liquid. As the lungs inflate, the alveoli increase in radius; as the lungs deflate, the radius diminishes. Because of the weight of the lungs and differences in transpulmonary pressure across the alveoli in varying regions of the lungs (see Chapter 4), alveoli do not increase and decrease in size uniformly throughout the lungs over the course of the respiratory cycle. During exhalation, the smallest alveoli, which tend to be in the more dependent or lower regions of the lungs, in a gravitational sense, are most prone to collapse.

 THOUGHT QUESTION 3-6: What would happen to multiple bubbles (each with an air–liquid interface) of different sizes that are connected?

The surface tension of water is constant across the entire surface and not dependent on the area of the air–liquid interface. Thus, in keeping with the law of LaPlace, bubbles with a small radius must have a greater pressure within them than bubbles with a large radius (Fig. 3-12).

 Because gas flows from regions of high pressure to regions of lower pressure, the small bubbles empty into the larger bubbles. Use Animated Figure 3-12 to view how bubbles with different sizes and pressures evolve over time when connected together. If the alveoli behaved as soap bubbles do, small alveoli would collapse into larger alveoli, and gas exchange, (i.e., the ability to get oxygen into the blood and to remove carbon dioxide from the blood) would be seriously deranged. Given the variable sizes of alveoli and their

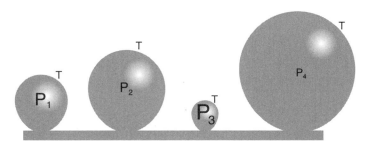

FIGURE 3-12 Bubble experiment. Four soap bubbles that share a common surface are shown. Note that the surface tension is the same for all bubbles. By LaPlace's law (P = 2T/r), we know that the pressure inside the bubble varies inversely with the radius. Thus, small bubbles have a large intra-bubble pressure. What do you think will happen to these bubbles given the different pressures within them? **See Animated Figure 3-12.**

interconnectedness via the airways, something must reduce the surface tension forces as alveoli decrease in size during exhalation, or segments of our lungs would collapse repeatedly with each breath. That "something" provided by the type II pneumocyte is a detergent called **surfactant**.

Surfactant

Detergents are substances with a polar and a nonpolar end. The molecules of a detergent are interposed between the polar molecules of water. Recall from our discussion of surface tension that the molecules of a liquid at an air–liquid interface are affected by particularly strong intermolecular forces. The interposition of a detergent between the molecules at the surface layer diminishes those forces and, thereby, reduces surface tension. Typically, a detergent's effect on surface tension is constant regardless of the surface area into which it is dispersed.

Surfactant, known chemically as dipalmitoylphosphatidyl choline, has the lowest surface tension of any biological substance ever measured. The type II pneumocytes lining the lung produce surfactant, which enters into the liquid lining the alveolus. Unlike most detergents, surfactant's effect is variable at different surface areas, and it is further dependent on the direction of change in the surface area (i.e., whether the alveolus is being inflated or deflated).

During lung inflation, molecules of surfactant appear to move into the layer of molecules at the surface of the liquid lining the alveolus (i.e., at the air–liquid interface). This process reduces surface tension and makes it easier to expand the lung. In this sense, surface tension plays a role, along with the elasticity of the lung tissue, in determining the compliance of the lung. As surface tension is reduced, compliance increases. As lung volumes further increase, the density of surfactant molecules in the surface layer remains constant, and the elastic properties of the lung reassert themselves as the main determinant of compliance.

During deflation of the lung, the density of surfactant molecules increases as surface area declines. This leads to a decrease in surface tension. If surface tension declines, pressure in the alveolus may decrease without leading to a significant reduction in the size of the alveolus (recall the law of LaPlace). With further exhalation, the alveolar radius diminishes, and some of the surfactant molecules are extruded from the surface layer to form subsurface collections, or **micelles**, while others are degraded. The molecules in the micelles may be recruited to enter the surface layer during the subsequent inhalation. The density of surfactant molecules in the surface layer during this second phase of exhalation appears to be constant. As the size of the alveoli and the surface area continue to decrease during the final stage of exhalation, surfactant density reaches a maximum. Further decreases in radius of the alveolus are not associated with greater decreases in surface tension.

Viewing the lung in its entirety, the effect of surfactant can be viewed as *increasing pulmonary compliance*, that is, the change in pressure required to achieve a given change in lung volume is less in the presence of surfactant than if it did not exist. In addition, during exhalation, surfactant, by reducing surface tension, helps to prevent alveolar collapse and derangements in gas exchange that result when alveoli have no air within them.

A third effect of surfactant is to minimize transudation of fluid from the pulmonary capillaries that line the alveoli. The forces associated with surface tension have the effect of reducing the hydrostatic pressure in the tissue around the capillaries; in a sense, the surface tension tends to "suck" fluid from the capillaries and facilitates accumulation of fluid in the alveolar space. Thus, the reduction in surface tension associated with the presence of surfactant helps to keep the alveoli dry.

THOUGHT QUESTION 3-7: What would happen in a disease state in which there is no surfactant? How would compliance change? What would happen to the alveoli?

Hysteresis

Consider the following experiment. We take a sheep lung in vitro, fill it with liquid, and inflate it and deflate it by moving liquid into and out of the lung. Now we remove the liquid (at least most of the liquid from the airways and alveoli, leaving only a thin liquid layer within the alveoli themselves) and inflate and deflate the lung by moving air into and out of the lung. Plot the pressure–volume curves for each condition. Will they look the same or different?

As you can see in Figure 3-13, when the lung is filled with liquid, the inflation and deflation curves can be superimposed. In contrast, when the lung is inflated with air, the two curves become separated; the inflation curve is shifted to the right of the deflation curve. This characteristic of air-filled lungs, the separation of the inflation and deflation limbs of the pressure–volume curves, is known as **hysteresis**. Because the only difference between the two conditions is the creation of an air–liquid interface, we hypothesize that hysteresis is a reflection of surface forces that accompany such an interface (i.e., surface tension) and the role of surfactant in modulating surface tension.

At the alveolar level, we discussed how surfactant molecules, newly produced by type II pneumocytes or drawn from subsurface micelles, enter into the liquid surface layer after the initiation of inhalation and begin to reduce surface tension. Thus, the initial part of the inflation curve is relatively flat; surface tension is high and compliance is low. As the density of surfactant in the surface layer increases, surface tension decreases and compliance increases; the slope of the curve becomes steeper. Eventually, as the volume of the alveolus

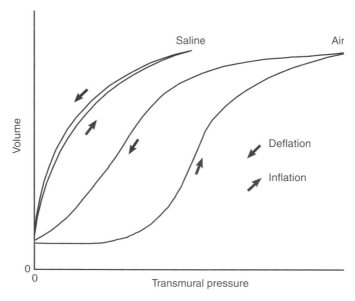

FIGURE 3-13 Hysteresis. The passive pressure–volume relationship is shown for a lung filled with liquid and one expanded with air. Note that the inspiratory and expiratory curves are superimposed when the lung is filled with liquid, but they are separated when the lung is expanded with air. This separation of the inspiratory and expiratory curves is called **hysteresis** and is thought to reflect the differential effect of surface forces (air–liquid interface) during the two phases of the respiratory cycle.

continues to increase, surfactant density remains constant and the elastic recoil forces of the lung become the primary explanation for changes in compliance. At high lung volumes, the elastic recoil of the lung increases, compliance decreases, and the slope of the curve begins to flatten.

During deflation, the density of surfactant molecules in the surface layer rapidly increases; surface tension decreases, and the initial portions of the deflation curve are fairly flat. In essence, alveolar pressure decreases, but because of the concomitant decrease in surface tension, there is little change in volume. With further deflation, some surfactant molecules are extruded from the surface layer, and the density of surfactant remains relatively constant. The compliance of the lung, reflecting effects of elastic recoil and surface tension, is also constant during this phase of exhalation. Surfactant appears to have a greater effect on compliance of the lung during exhalation than during inhalation, probably because of the time course for surfactant molecules to enter and exit the surface layer. As the lung is expanding, there is some delay in the movement of molecules of surfactant from the subsurface to the surface. Thus, for any given lung volume, the compliance of the lung is less during inhalation than during exhalation. If we view the pressure–volume curves and hysteresis from the perspective of the entire lung rather than at the level of the alveolus, another factor must be considered. At the end of exhalation, even with surfactant present, some alveoli collapse, particularly at the bases of the lung, where transpulmonary pressure may be negative (pressure outside the alveolus greater than pressure inside the alveolus). To reopen these collapsed alveoli at the beginning of the next inspiration, or *recruit* them to participate in gas exchange (remember that a collapsed, airless alveolus cannot provide oxygen to or eliminate carbon dioxide from the pulmonary capillaries), requires a relatively high change in pressure. This pressure, termed the *critical opening pressure*, may be as high as 20 to 30 cm H_2O in animal experiments. Thus, the initial portion of the inflation curve, with its fairly low slope, may partly reflect the recruitment of collapsed alveoli.

One other factor may contribute to the observation of hysteresis in the lungs. When one inflates the lung and holds the lung at the higher volume for several seconds, the elastic recoil forces appear to diminish slightly, a phenomenon called *stress relaxation*. Conversely, after deflation of the lung, the recoil forces increase; this process is called *stress recovery*. This phenomenon, called **stress adaptation**, is believed to be caused by the viscoelastic properties of smooth muscle and connective tissue in the lung but, relative to the effect of surface forces discussed above, plays a small role in hysteresis.

PUTTING IT TOGETHER

A 30-year-old, severely obese woman comes to see you. She indicates that she has trouble breathing when she lies flat but feels better within a few seconds if she stands up. Her oxygen and carbon dioxide levels are normal, as is her chest radiograph, and she has no history of lung or heart disease. Measurements of her pulmonary function tests indicate that her TLC, FRC, and RV are all reduced, although the RV is not reduced as much as TLC and FRC. The strength of her inspiratory and expiratory muscles is normal. What is the likely physiologic cause of her breathing discomfort?

A patient with severe obesity has reduced chest wall compliance. Remember that the abdomen forms the inferior portion of the chest wall. A large abdomen impairs the movement of the diaphragm during inspiration; therefore, a greater change in pressure is required to achieve a given change in volume. This additional burden imposed on the ventilatory muscles may be perceived as an increase in the work or effort of breathing. The TLC and FRC are reduced because the balance of forces that determine these volumes has been altered; there is now a greater deflationary or collapsing pressure relative to the normal state because of the weight of the chest wall. The RV may be relatively preserved or even increased because of the tendency for airways to collapse at the bases of the lungs during exhalation; collapse of airways before alveolar air has exited is termed air trapping. The weight of the chest wall in an obese individual may make pleural pressure positive at the base of the lungs, thereby creating a negative transmural pressure across the airway towards the end of exhalation; negative transmural pressure is a collapsing pressure.

Summary Points

- The study of the respiratory system during conditions in which there is no gas flow is termed *statics*.
- Both the chest wall and the lungs possess elastic properties.
- The chest wall and lungs have a resting or unstressed volume. To change the volume from the unstressed level requires force. After release of the force, the chest wall and lungs return to their resting volumes.
- The major lung volumes—TLC, FRC, and RV—are determined primarily by the balance of forces exerted by the chest wall, lungs, and ventilatory muscles.
- The pleura plays a crucial role in linking the forces exerted by the lungs and chest wall.
- The transmural pressure is the pressure across a wall of an enclosed structure. The transmural pressure is the pressure inside the structure minus the pressure outside the structure. Positive transmural pressures are associated with distending forces that increase the volume of a structure. Negative transmural pressures are associated with collapsing forces that decrease the volume of a structure.
- The TLC is the volume of the lung at the end of a maximal inspiration.
- The RV is the volume of the lung at the end of a maximal expiration.
- The FRC is the volume of the lung at the end of a relaxed expiration.
- In older adults, airway closure is a contributing factor in determining the RV.
- The tension generated by a muscle is related to the length of the muscle at the time that it is stimulated by a neurological impulse. The longer the muscle at the time of neurological stimulation, the greater the tension generated.
- The diaphragm is the major inspiratory muscle.
- Compliance is a characteristic of a structure that relates the transmural pressure change required to achieve a given change in volume.
- The compliance of the respiratory system primarily depends on the elastic properties of the lungs and chest wall in healthy individuals.
- The volume of the lung depends on the compliance of the respiratory system and the transpulmonary pressure.
- FRC can be measured with the helium dilution or nitrogen washout techniques. Both of these tests may underestimate FRC in people with airway disease that impairs communication of regions of the lungs with the central airways. Measurement of FRC with a body plethysmograph is less affected by airways disease.
- The action of surfactant to reduce surface tension in the alveolus is critical in increasing respiratory system compliance and in minimizing alveolar collapse during exhalation.
- Surface forces are the primary cause of the hysteresis seen in the pressure–volume curves of the lungs during the respiratory cycle.

Answers TO THOUGHT QUESTIONS

3-1. When the pleural space is disrupted, as occurs when air enters the pleural space because of a hole in the lung (pneumothorax), the relationship between movement of the chest wall and the lung is eliminated. There is no longer a vacuum between the chest wall and the lung. The natural elastic properties of the lung cause the lung to collapse. In an otherwise healthy lung with normal airways, the volume of the collapsed lung can be very small, approximating the tissue volume of the lung itself (i.e., there is very little gas left in the lung). Use Animated Figure 3-1A to show the changes in the lung and chest wall when air is traumatically introduced into the pleural space. Note how small the lung can become, close to its equilibrium position or minimal volume, as well as how the chest wall expands on the affected side.

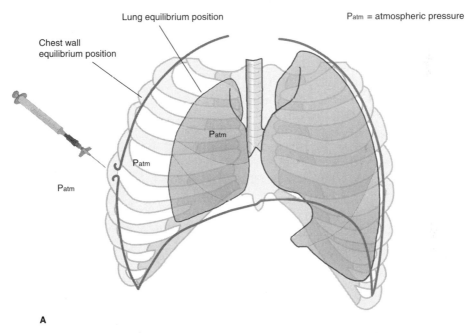

A

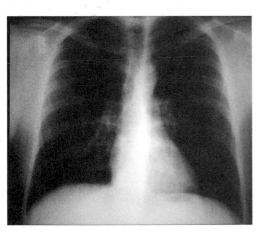

Normal Chest X-Ray

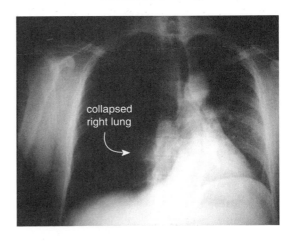

Pneumothorax X-Ray

B

3-2. The size of the balloon is determined by the transmural pressure (pressure inside the balloon minus the pressure outside the balloon, or $P_{TM} = P_{inside} - P_{outside}$; recall that pressure outside the balloon initially is atmospheric which, by convention, is taken as 0 at sea level). When the balloon is placed under the water, the pressure outside the balloon increases. This reduces the transmural pressure, thereby reducing the size of the balloon. Use Animated Figure 3-2A to drag a balloon underwater and observe the changes in transmural pressure and volume.

As the gas in the balloon is compressed, the pressure in the balloon increases. Thus, under water, the balloon comes to a new equilibrium with a smaller size despite the fact that the pressure inside the balloon is greater than it was when it was above the surface of the water. The most common misconception is that only the intracavitary (or pressure inside) a structure matters in determining its volume. As we can see from this example, it is the pressure across the wall, or transmural pressure, that matters.

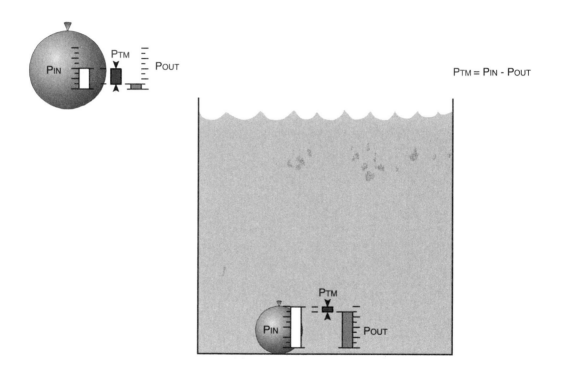

3-3. Compliance can be calculated as the change in volume attained for a given change in pressure. In Figure 3-6, the x axis represents transmural pressure, and the y axis is the volume of lung starting at RV. As read from the graph in Figure 3-6, in the normal lung, there is approximately a 4 L difference between RV and TLC; to make this change in volume, one must have a transpulmonary pressure of approximately 30 cm H_2O. Thus, the compliance of the normal lung is 4000 mL/30 cm H_2O or 133 mL/cm H_2O.

The calculations below for the diseased lungs are done similarly by estimating the pressure and volume ranges from the graph.

The compliance of a lung in an individual with emphysema is 4500 mL/15 cm H_2O or 300 mL/cm H_2O. The compliance of the fibrotic lung is 2500 mL/40 cm H_2O or 63 mL/cm H_2O.

3-4. When you encounter a patient with reduced lung volumes, you must analyze the situation with an eye toward determining whether the abnormality is caused by a problem with the lungs, the chest wall, or the muscles of ventilation. The FRC is determined by the elastic recoil inward of the lungs and the tendency, at that lung volume, for the chest wall to spring outward. FRC, the volume of the respiratory system at the end of a passive exhalation, does not require the activity of muscles. Thus, if the FRC is normal, it suggests that the elastic properties of the lung and chest wall are normal. In this setting, a reduced TLC is most likely caused by inspiratory (e.g., diaphragmatic) muscle weakness. If the RV is normal, the expiratory muscles can be assumed to be functioning normally.

3-5. In a patient who has regions of lung that communicate poorly with the central airways (e.g., a patient with severe emphysema), the helium may not diffuse well into the poorly ventilated regions of lung. Thus, from the standpoint of the helium dilution technique, it is as if these regions of lung do not exist. One may get a lung volume measurement in such cases that underestimates the true lung volume by a significant amount.

3-6. Assuming that there was no surface active agent to reduce surface tension as a bubble gets smaller, the surface tension in all the communicating bubbles is the same even though they have different sizes. In the smaller bubbles (i.e., the ones with a smaller radius), the pressure inside the bubble must be bigger than in the large bubbles (by virtue of LaPlace's law). Consequently, there will be a pressure gradient for air to flow from the small bubbles into the large bubbles. The small bubbles will collapse, ultimately leaving you with one large bubble. Press the play button in Animated Figure 3-12 to allow air to flow between the bubbles and watch how they evolve over time. Imagine if alveoli operated that way; it would not be very effective for gas exchange.

3-7. In a disease in which there is no surfactant, alveoli tend to collapse as they get smaller during exhalation (remember the soap bubbles in Thought Question 3-6). This results in large regions of "shunt" in the lung (blood flow traveling to alveoli in which there was no gas entering the alveolus), which produces significant hypoxemia (see Chapter 6). In addition, when one tries to expand the lung on the next breath, it would require a large transpulmonary pressure to get the desired change in lung volume because surface tension is very high. Consequently, the effective compliance of the lung is reduced. Two disease states characterized by absent or malfunctioning surfactant are the infant respiratory distress syndrome (also known as hyaline membrane disease) and the adult respiratory distress syndrome (ARDS). The former is seen in premature infants whose lungs have not matured to the point that they can produce surfactant. ARDS is a condition that is associated with a wide range of disorders (e.g., pneumonia, sepsis, smoke inhalation) that cause damage to the alveolar–capillary interface.

Review Questions

DIRECTIONS: *Each of the numbered items or incomplete statements in this section is followed by answers or by completions of the statement. Select the ONE lettered answer or completion that is BEST in each case.*

1. A patient with symmetrical reductions in TLC, FRC, and RV is more likely to have:

    A. inspiratory muscle weakness
    B. expiratory muscle weakness
    C. generalized muscle weakness
    D. a condition that reduces compliance of the lungs

2. A 22 year-old medical student decides to go SCUBA diving. He descends to a depth of 50 feet, takes a breath, and then holds his breath while ascending to the surface. The lung volume at the surface compared with his lung volume at a depth of 50 feet is:

    A. smaller
    B. larger
    C. the same

3. In a patient with severe emphysema, there is a reduction in the elastic recoil of the lungs because of the destruction of the lung tissue associated with this disease. This is associated with each of the following findings except:

    A. an increased TLC
    B. an increased FRC
    C. a lengthened diaphragm at the beginning of inspiration
    D. a diminished ability to generate tension in the diaphragm at the beginning of inspiration

4. Assuming the individual is holding his breath with the glottis open in both conditions, the transpulmonary pressure at TLC compared with FRC is:

    A. greater
    B. smaller
    C. the same

5. A patient has adult respiratory distress syndrome (ARDS), a condition that alters the alveolar–capillary interface and the function of surfactant. The patient is in respiratory failure and is being sustained on a ventilator. You have the option with the ventilator to keep his FRC at different volumes. Using what you know about the Law of LaPlace and the factors that determine alveolar stability, you decide to improve his ability to get oxygen into his lungs by maintaining FRC at:

    A. the same volume as his normal FRC
    B. a volume below normal FRC
    C. a volume above normal FRC

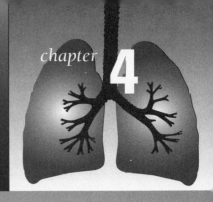

chapter 4

Dynamics: Setting the System in Motion

CHAPTER OUTLINE

FLOW THROUGH TUBES: BASIC PRINCIPLES
- Resistance
- Laminar Versus Turbulent Flow
- Airway Size and the Distribution of Resistance in the Lungs
- Bernoulli's Principle

FLOW THROUGH TUBES: APPLICATION TO THE LUNGS
- Lung Resistance
- Determinants of the Cross-sectional Area of the Airway
- Transmural Pressure and the Size of Airways During Forced Exhalation

THE FLOW–VOLUME LOOP AND FLOW LIMITATION
- The Volume–Time Plot
- Maximal Expiratory Flow: The Forces to Be Considered
- Expiratory Flow Limitation
- Presence of Flow Limitation on the Flow–Volume Loop
- Time Constants

PUTTING IT TOGETHER

SUMMARY POINTS

LEARNING OBJECTIVES

- **To differentiate the characteristics of the respiratory system when there is air flow and change in volume (dynamics) from the respiratory system in the absence of air flow (statics).**
- **To describe the role of the different elements of the ventilatory pump when the respiratory system is placed in motion.**
- **To define the principles associated with the flow of gas through tubes.**
- **To delineate the factors that contribute to airway and tissue resistance in the lungs.**
- **To define the concept of flow limitation and the principles that lead to this phenomenon.**
- **To define and integrate the physiological principles that are necessary to understand the flow–volume loop.**

Having read about the characteristics of the respiratory system under static conditions, you are now ready to examine what happens when we set the system in motion—that is, when we breathe in and out. We continue to focus on the ventilatory pump in this chapter. Chapter 3 outlined how the elements of the pump work together to generate a negative alveolar pressure that serves as the driving force for the movement of gas into the lungs. At the end of an inhalation, the pump, during resting breathing, relaxes. The recoil forces of the lungs (and of the chest wall at higher lung volumes) cause alveolar pressure to become positive so exhalation may follow and the system can be reset for the next breath. We will now examine what happens when the flow of air begins. We described the study of statics as "snapshots" of the respiratory system. **Dynamics**, the examination of the respiratory system during conditions of flow of gas into and out of the lungs, is the motion picture version of our story.

To begin the study of dynamics, we must examine a series of physical principles that characterize the flow of fluids through tubes. We will build these principles, one upon another, until we are able to construct the flow–volume loop, the most common graphical representation used by clinicians to understand the respiratory system in the dynamic state. A key to understanding the flow–volume loop and the physiology of obstructive lung diseases such as asthma and emphysema is the concept of flow limitation. By the time you complete this chapter, you will have a firm understanding of these principles and a glimpse of their application in the evaluation of patients with diseases of the airways.

Flow Through Tubes: Basic Principles

RESISTANCE

Whenever a fluid flows through a tube, it encounters resistance. To create and sustain flow, a **driving pressure** must be present to provide the energy to move the fluid and to overcome resistive forces. Think back for a moment to your study of electricity. Ohm's law governed the flow of electrons in a circuit.

$$\text{Potential difference} = \text{Current} \times \text{Resistance}$$
$$V = I \times R$$

When considering the flow of gas in a tube, one can make an analogy to the electrical circuit. The potential difference is the driving pressure, that is, the pressure change from one end of the tube to the other. The current in the circuit is analogous to the flow in the tube, and the electrical resistance of the circuit is similar to the resistance of the tube to flow. Thus, Ohm's law is transformed into the following relationship that describes flow through tubes.

$$\text{Change in pressure} = \text{Flow} \times \text{Resistance}$$
$$\Delta P = \dot{V} \times R$$

Note: In respiratory physiology, we indicate flow by writing a V with a dot over it, $\dot{V}$, indicating volume per time (Table 4-1).

To double the flow for a given resistance, one must double the driving pressure. Viewed from another perspective, a doubling of resistance causes a doubling of the pressure decrease across the length of the tube for any given flow.

LAMINAR VERSUS TURBULENT FLOW

As the molecules of a fluid travel through a tube, they may be aligned in a variety of ways relative to the direction of flow. The particular alignment makes it easier or more difficult

TABLE 4-1 Ohm's Law and Flow Through Tubes	
OHM'S LAW	**FLOW THROUGH TUBES**
Potential difference	Driving pressure
Current	Flow
Resistance	Resistance
$V = I \times R$	$\Delta P = \dot{V} \times R$

I = current; R = resistance; ΔP = change in pressure; $\dot{V}$ = flow; V = potential difference.

for the fluid to flow through the tube or, stated differently, depending on the alignment of the molecules' movements, different amounts of energy are needed to maintain a given flow.

In **laminar flow**, all of the molecules are moving in a direction that is parallel to the long axis of the tube. Fluid traveling in or near the center of the tube moves faster, that is, it has a greater velocity, than fluid near the wall of the tube. Under conditions of laminar flow, the change in pressure between two points within the tube is proportional to the flow of the fluid within the tube.

$$\Delta \text{Pressure} \propto \text{Flow}$$
$$\Delta P \propto \dot{V}$$

In contrast, under conditions of **turbulent flow**, some of the molecules in the fluid travel parallel to the long axis of the tube and other molecules travel in a variety of other directions, including perpendicular to the long axis.

THOUGHT QUESTION 4-1: **What sound do you hear in patients with lung problems that cause increased turbulence in the airways? Why?**

To achieve a given flow when turbulent conditions are present, one must have a greater driving pressure. With turbulent flow, the change in pressure between two points in the tube is proportional to the square of the flow.

$$\Delta \text{Pressure} \propto (\text{Flow})^2$$
$$\Delta P \propto (\dot{V})^2$$

If there is a greater pressure decrease over the same length of the tube to achieve a given flow, this means more energy is needed (compared with a smaller pressure decrease) to maintain that flow (Fig. 4-1).

During resting breathing, flow is turbulent in the trachea and laminar in the small peripheral airways. At branch points, one often sees a type of flow that manifests some characteristics of both laminar and turbulent flow. This is called marginally **laminar or transitional flow** because it occurs at transitions between laminar and turbulent flow. In these locations, flow is generally laminar, but there are also eddies generated at angles to the long axis of the tube.

Whether a tube through which a gas is flowing has laminar or turbulent flow depends on a number of factors, including the viscosity of the gas, the density of the gas, the radius of the tube, and the velocity of the molecules in the tube (note: velocity, which has units of distance/time is not the same as flow, which has units of volume/time). The impact of these factors on the type of flow is expressed by Reynolds number.

$$\text{Reynolds number} = \frac{2(\text{Radius})(\text{Density of the gas})(\text{Velocity of the gas})}{\text{Viscosity of the gas}}$$

The larger Reynolds number is, the greater the likelihood that flow will be turbulent. For example, flow is likely to be turbulent if the radius of the tube is large and the velocity of the gas is high. In a rapidly branching system such as the lung, fully laminar flow probably only occurs in the small airways. The very low velocity of the gas in the small airways near the periphery of the lung and the small radii of the individual airways in the periphery contribute to conditions that produce laminar flow.

Movement of molecules

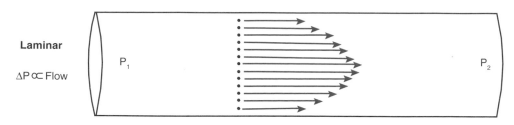

Laminar

$\Delta P \propto Flow$

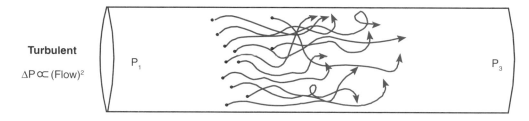

Turbulent

$\Delta P \propto (Flow)^2$

For same flow, $P_2 > P_3$

FIGURE 4-1 Energetics of laminar and turbulent flow. Under conditions of laminar flow, the molecules of the fluid travel in a direction that is parallel to the long axis of the tube. In contrast, turbulent flow is characterized by molecules traveling in a variety of directions, including perpendicular to the long axis of the tube. Turbulent flow may be associated with vibration of the walls of the tube. Under conditions of laminar flow, the decrease in pressure from one end of the tube to the other is proportional to the flow. If turbulent flow is present, however, the decrease in pressure is proportional to the square of the flow. Thus, for a given flow, pressure decreases to a greater degree when turbulence is present than when one has laminar flow ($P_2 > P_3$). Stated in another way, more energy is required to maintain a given flow under turbulent conditions.

QUICK CHECK 4-1   FACTORS THAT DETERMINE LAMINAR VERSUS TURBULENT FLOW

Radius of the tube
Density of the gas
Velocity of the gas
Viscosity of the gas

In a healthy individual, most turbulent flow is found in the central airways, particularly the trachea. During exercise, when the velocity of the gas increases as ventilation increases, turbulent flow extends more deeply into the lung, perhaps as far as the fifth or sixth generation of airways. Between the central and the peripheral airways, transitional flow is present.

THOUGHT QUESTION 4-2: A patient comes to the emergency department with shortness of breath caused by spasm of the vocal cords (laryngospasm). In this condition, the narrowing of the airway and consequent increase in velocity of gas moving across the vocal cords result in worsening turbulent flow. What intervention could you make to reduce the work of breathing and ease the patient's symptoms while waiting for the treatment of the laryngospasm to take effect?

As you take care of patients in the clinical setting, you will never have to calculate Reynolds number. The case described in Thought Question 4-2, however, demonstrates the importance of the principles underlying turbulent and laminar flow.

AIRWAY SIZE AND THE DISTRIBUTION OF RESISTANCE IN THE LUNGS

As we continue to examine the factors that determine resistance in the airways, we must now consider another principle of physics, Poiseuille's law. According to this law, the flow of gas, under laminar conditions, is proportional to the change in pressure and the radius of the tube to the fourth power.

$$\dot{V} = \frac{\Delta Pressure\ (\pi)\ r^4}{8(Viscosity)(Length)}$$

Recall our formula that states $\Delta Pressure = Flow \times Resistance$ (or restating, Flow = $\Delta Pressure$ divided by resistance).

$$\dot{V} = \frac{\Delta Pressure}{Resistance}$$

Resistance, which varies inversely with flow, is thus proportional to the length of the tube and inversely proportional to the radius of the tube to the fourth power.

$$Resistance = \frac{8\ (Viscosity)(Length)}{\pi\ (r^4)}$$

$$Resistance \propto \frac{1}{r^4}$$

Let us consider the implications of this relationship. All other things being equal, a longer tube has a greater resistance than a shorter tube, and if one reduces the radius of a tube by one half, the resistance increases by a factor of 16!

Where is resistance highest in the lung? As air moves from the mouth to the alveoli, the radius of the airways gets smaller and smaller. Given the principles of Poiseuille's law, you might conclude that the highest resistance in the lungs resides in the periphery. Another important concept must be considered, however, before you can answer this question. As you travel from the trachea to the alveolus, you witness the branching of succeeding generations of airways (see Chapter 2). Each single airway branches into two or more *parallel* airways with a combined cross-sectional area greater than the parent airway. In a sense, the radius of the combined smaller airways is greater than the radius of the larger parent airway; thus, the resistance is actually less.

As you put tubes together in various combinations and consider the resulting resistance, you must pay particular attention to the arrangement of the tubes. Are they in series or in parallel? For tubes in series, the total resistance is equal to the sum of the resistances of the individual tubes:

$$R_{tot} = R_1 + R_2 + R_3$$

The total resistance is greater than the resistance of any individual tube.

In contrast, for airways in parallel, the total resistance is less than the resistance of any of the individual tubes:

$$1/R_{tot} = 1/R_1 + 1/R_2 + 1/R_3$$

As you move from the mouth to the alveoli, you go from a single airway (the trachea) to millions of airways (terminal and respiratory bronchioles) in parallel. Consequently, the resistance of the lungs is much less in the small airways in the periphery than in the central airways. Within the more central airways, one must also consider the effect of the smooth muscle in the medium-sized bronchi. The radius of these airways may vary considerably depending on the tone of the muscles and whether they are actively contracting as a result of a change in the balance of the sympathetic and parasympathetic nervous systems and disease states such as asthma.

Under normal conditions, the bulk of resistance in the lung resides in the first six to seven generations of airways (Fig. 4-2). Experimental data indicate that the peak airway resistance also occurs in the fifth to seventh generation of airways. The reason or reasons for this is not entirely clear, but it may be the consequence of the geometry of the airways at the initial branchpoints in the trachobronchial tree. As emphasized previously, the cross-sectional area

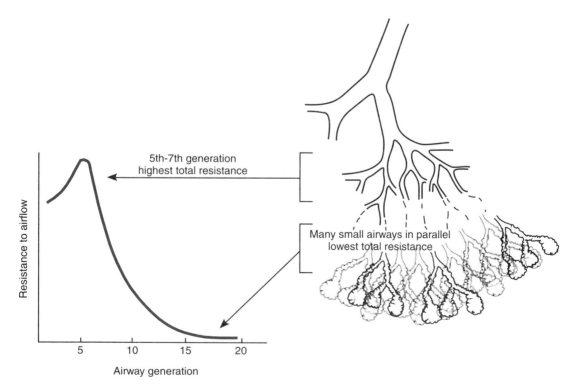

FIGURE 4-2 Distribution of resistance in the lung. Although the smallest airways, as individual tubes, have the highest resistance, they are arranged in parallel. This arrangement means that the resistances of the small bronchi add as reciprocals, thereby leading to extremely low resistance to airflow in the periphery of the lung. The highest areas of resistance are found in the more central airways, with the peak airway resistance occurring in the fifth to seventh generation of airways.

of the airways in parallel gradually increases as one goes from the trachea to the periphery of the lung, a finding that accounts for the fact that most of the resistance in the lungs is attributable to the central airways. Diseases of the more central airways can, therefore, have a significant impact on total resistance in the lung and, consequently, can be easily detected by tests that assess flow or resistance. In contrast, diseases of the small airways have relatively little impact on lung resistance and can be more difficult to detect with standard tests of pulmonary function. For this reason, the region of the lungs containing the small airways has been called the *silent zone* of the lungs.

Let us revisit the concepts of laminar and turbulent flow in the context of changing airway size. Where are you more likely to encounter turbulent flow, in the central or the peripheral airways? Consider what happens as gas is exhaled from the alveoli. As you go from the millions of airways in the periphery to a diminishing cross-sectional area in the central airways, the velocity of the gas must increase because flow is constant. Imagine four single-lane roads arranged in parallel that are now merging into a two-lane highway. Assume that the flow of cars (number of cars/min) is constant from the single-lane roads to the two-lane highway. In order to accommodate all those cars in the two-lane highway and maintain the flow, the velocity of each car must increase (Fig. 4-3).

 Use Animated Figure 4-3 to select either two or four single-lane roads merging and compare the flow and velocity (shown as speed) of the cars for these scenarios. Notice in

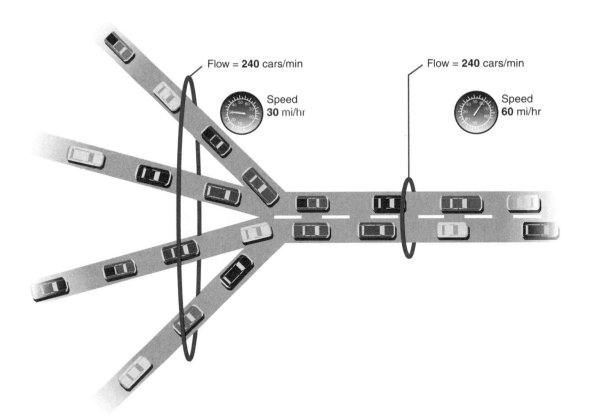

FIGURE 4-3 Change in velocity—the airways as roads. Four single-lane roads merge into a two-lane highway. The flow of cars (number of cars/min) on the four roads must be the same as the flow on the two-lane highway. For flow to be constant, the velocity (shown as speed in miles/hour) of the cars must increase when moving onto the two-lane highway. This same principle is true of gas as it moves from the small airways in the periphery of the lung to the trachea during an exhalation. **Also see Animated Figure 4-3.**

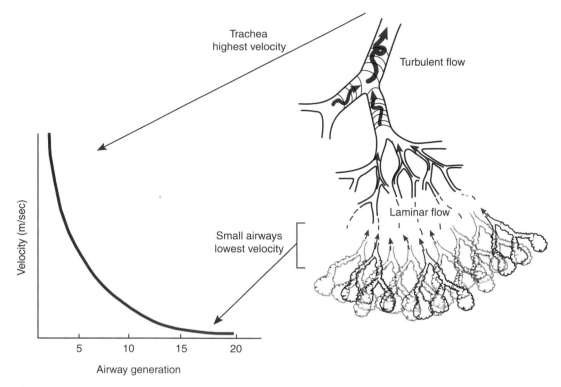

FIGURE 4-4 Distribution of velocity of gas in the lung. As gas moves from the periphery of the lungs to the central airways and from millions of airways in parallel to a single trachea, it picks up speed. Velocity is highest in the trachea. Because velocity plays an important role in determining whether flow is turbulent or laminar, turbulence is more likely in the central airways.

particular how the cars must speed up to maintain flow in the case where four single-lane roads are merging into two. As already discussed, this is analogous to the movement of gas during exhalation from the many small airways in the periphery of the lung to the lesser cross-sectional area of the central airways. Before reading on, think about how this change in velocity of gas affects the type of flow (laminar versus turbulent) in the peripheral and central airways.

According to Reynolds number, gas moving through a tube at a higher velocity is more likely to demonstrate turbulent flow. Thus, whereas flow in the central airways is more likely to be turbulent, flow in the small bronchioles is laminar (Fig. 4-4).

During normal resting breathing (at lung volumes approximating functional residual capacity), 80% of the resistance to flow is in the airways greater than 2 mm in diameter, and nearly half of the total resistance of the lungs is in the central airways.

THOUGHT QUESTION 4-3:   When an individual engages in aerobic exercise and increases ventilation, what does he or she do to reduce resistance to airflow and thereby decrease the work of breathing? (Hint: look at the person's face.)

BERNOULLI'S PRINCIPLE

What makes airplanes fly? As the thrust from the engines propels the airplane down the runway, air travels above and below the wing. An airplane wing is shaped so that the distance across the surface from the front to the back of the wing is longer on the top than on the bottom of the wing. Because the flow of air must be the same above and below the wing, the velocity of the gas traveling over the top of the wing must be greater than that beneath the wing. The increased velocity results in a lesser pressure exerted by the gas. In other words, the air above the wing exerts less pressure on the wing than does the air below the wing; the result is "lift," and the plane rises off the runway. This relationship between the velocity of a gas and pressure is Bernoulli's principle, and it is also applicable to the airways of the lungs, as we will see shortly (Fig. 4-5).

First, let us apply Bernoulli's principle to fluids flowing through tubes. Imagine a fluid moving at a given flow through a tube with a wide diameter. The molecules in the fluid have a certain velocity and exert a pressure within the tube. The tube abruptly narrows. If we stipulate that the flow remains constant throughout the system, the velocity of the molecules of the fluid must now increase. As with the example of the airplane wing, to move the molecules at a greater velocity, keeping flow constant, requires work to impart energy to the molecules. This can only occur if there is greater pressure in the wider part of the tube. The difference in pressure between the wide and narrow sections of the tube reflects the force necessary to increase the velocity of the molecules. The greater the increase in velocity, the greater the decrease in pressure (Fig. 4-6).

If an external pressure is exerted against the outer walls of the tubes, the pressure inside the small tube may now be less than the pressure outside the tube. This principle plays an important role in the tendency of airways to collapse during a forced exhalation.

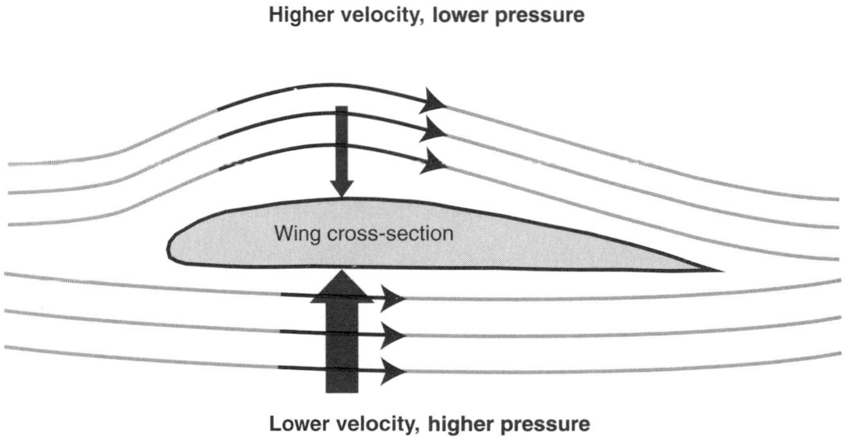

Higher velocity, lower pressure

Lower velocity, higher pressure

FIGURE 4-5 Bernoulli's principle and airplanes. An airplane wing cut in cross-section is shown. Notice that the distance across the surface of the wing from the front to the back is greater over the top than the bottom of the wing. Because the flow of gas above and beneath the wing must be the same, the velocity of the gas on top must be greater than that under the wing. Consequently, the pressure exerted by the gas down on the wing from above is less than the pressure exerted up on the wing from below. The result is a pressure differential that lifts the airplane off the runway.

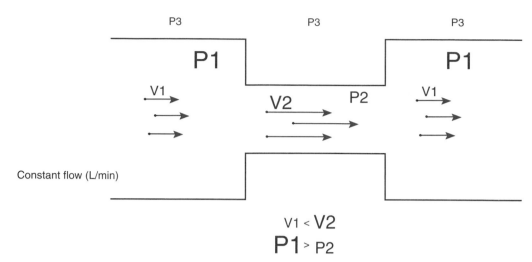

FIGURE 4-6 Bernoulli's principle and flow through tubes. Two tubes of equal diameter are connected by a smaller tube. The flow of gas through the tubes is constant throughout the system. The gas in the large tubes is moving with a velocity, V1; the velocity of the gas in the smaller tube is V2. The gas in the larger tubes exerts a pressure against the inside wall of the tube, P1; the gas in the smaller tube exerts a pressure, P2. Surrounding the tubes, there is a pressure P3 that is exerted against the outer wall. To maintain a constant flow, the gas must increase velocity in the small tube relative to the large tube (V2 > V1). Increasing the velocity of the gas requires a pressure decrease, and as a result, P2 < P1. If P1 > P3 > P2, the small tube will be compressed. Note that the sizes of the pressure and velocity labels in the figure reflect their magnitudes.

Flow Through Tubes: Application to the Lungs

LUNG RESISTANCE

As you move air into the lungs, you have to overcome two types of resistance: airway resistance and tissue resistance. Airway resistance is determined by the principles we have just outlined for the general flow of air or liquid through tubes. Tissue resistance is a manifestation of the forces that must be overcome by virtue of moving molecules as you stretch the lungs. Airway resistance accounts for approximately 85% of the total resistance of the lungs.

If you move air rapidly into the lungs, that is, generate a high flow, a greater driving pressure is required to overcome the resistance of the lungs. Imagine the relationship between the volume of the lungs and pressure in the alveolus as you go from one volume (V_1) to another (V_2) under infinitely slow (also called **quasistatic**) conditions (essentially a series of static measurements) (Fig. 4-7). Because we are under quasistatic conditions, we can essentially ignore the effects of resistance (resistance is only important when there is flow), and the relationship between volume and pressure is a straight line. The slope of the curve reflects only the compliance of the lung (recall that compliance is the change in volume divided by the change in pressure needed to produce the change in volume). Now move the air into the lung and expand it from V_1 to V_2 rapidly (i.e., under conditions of high flow). How do you think the relationship between pressure and volume will change? (Fig. 4-8)

To inflate the lung rapidly, you must overcome the resistance of the airways and the lung tissue. Initially, energy must be expended to accelerate the gas and to overcome the inertial forces of the lung. Thus, pressure changes with little change in volume at the outset of the inflation. As higher flows are achieved, volume changes more rapidly than pressure, as the momentum of the gas carries it into the alveoli, until the desired volume is reached.

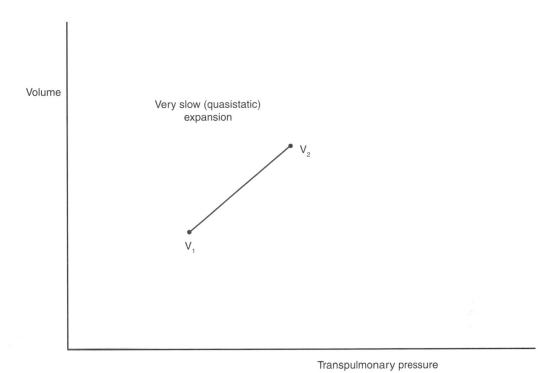

FIGURE 4-7 Expansion of the lung under quasistatic conditions. As you move from volume 1 (V_1) to volume 2 (V_2), the relationship between pressure and volume is linear. The slope of this line reflects the compliance of the lung (over the small volume range shown in this example, the compliance is constant). Because flow is essentially zero, resistance is negligible and does not enter significantly into the relationship between pressure and flow.

THOUGHT QUESTION 4-4: When you compare inflation of the lung under quasistatic conditions and under conditions of high flow, the pressures at the beginning of the inflation (P1) are the same under both conditions. Similarly, the pressures at the end of the inflation (P2) are the same under both conditions. Is the work to inflate the lung the same or different in these two circumstances? Why?

   Use Animated Figure 4-8 to observe how the speed of lung expansion affects the pressure–volume relationship and the amount of work involved in inflating the lung. First, choose a speed of lung expansion, and then press the Go button to run the scenario. You can also display the work on the graph as a shaded area. Compare very slow expansion (quasistatic conditions) to very fast expansion and note how the pressure volume curve changes shape and how the amount of work increases in relation to the speed of expansion.

DETERMINANTS OF THE CROSS-SECTIONAL AREA OF THE AIRWAY

As noted in Chapter 2, the larger, more central airways of the lung have cartilage and smooth muscle in their walls that helps determine their shape and size. The small airways in the periphery of the lung, however, have thin walls and are relatively compliant; their shape and size are affected to a greater degree by the pressure differential across the wall of the airway. Furthermore, the small airways are supported by a latticework of connective tissue (including the alveolar network) supplied by the surrounding lung. In a sense, the lung tethers open the small airways (Fig. 4-9).

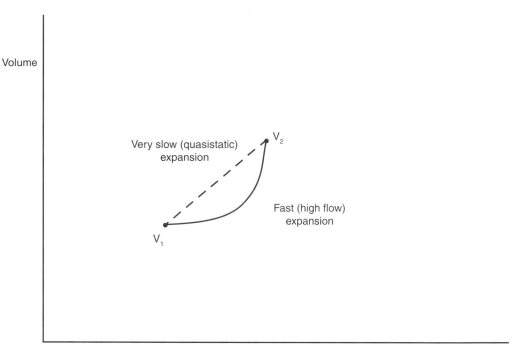

FIGURE 4-8 Expansion of the lungs under conditions of high flow. Notice how the curve linking volumes 1 and 2 (V_1 and V_2) is now different from Figure 4-7. At the beginning of the inflation, you must accelerate the gas and overcome the inertial forces present in the lung tissue. This is evident in the flat portion of the curve at the beginning of the infla-tion. Pressure is changing, but there is little change in volume. As soon as the flow has been increased, volume begins to change more quickly than pressure until you reach V_2. The momentum of the gas continues to fill the lung toward the end of inhalation even as the inspiratory muscles begin to relax. Thus, at the end of the fast inspiratory maneuver, the gas continues to fill the alveolus without a significant change in pressure. This is evident in the nearly vertical slope of the curve as you approach V_2. **Also see Animated Figure 4-8.**

Recall the example of the mesh stocking we discussed in Chapter 3. The network of the fibers forming the mesh is interconnected. Imagine a very flexible tube sewn into the mesh with the direction of the tube perpendicular to the plane of the mesh. As the mesh is stretched and the fibers pulled farther apart, the diameter of the tube passing through the mesh becomes larger. Similarly, as the lung expands, the diameter of the small airways in-creases; as the lung deflates, the caliber of the airway diminishes. Thus, lung size is one of the major determinants of the cross-sectional area of the small airways. Use Animated Figure 4-9 to view the change in diameter of a small airway during breathing. You can press the play button to watch the motion during inspiration and expiration or use the slider to control the animation yourself. Note how the small airway (shown in cross-sec-tion) runs through and is attached to the alveolar connective tissue network. Thus, during inspiration, when the lung inflates, expanding forces are transmitted from the pleura to the small airways via the intervening connective tissue.

The tethering effect just described can be related to the elastic properties of the lung tissue. If the elastic recoil of the lung is diminished and the tethering effect supplied by the lung is weakened, as occurs in emphysema, there will be a greater propensity for the small airways to narrow when the transmural pressure is negative (recall from Chapter 3 that transmural pressure is negative when the pressure outside the airway is greater than

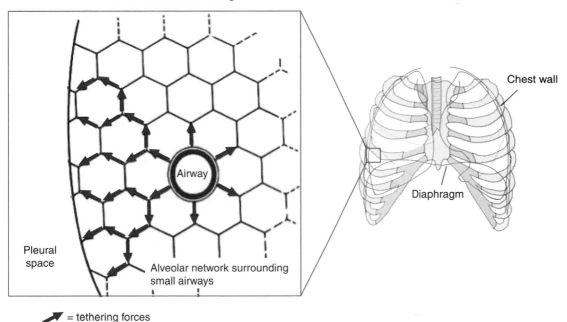

Cross-sectional cut of lung

= tethering forces

FIGURE 4-9 Small airways and the tethering effect of the lung. The small airways pass through lung tissue that forms a latticework support structure for the airways. As the lung inflates and increases in volume, the diameter of the airway increases. During deflation, the airway diameter decreases. The tethering effect, or elastic recoil force exerted by the lung, is greater at higher lung volumes. Also see Animated Figure 4-9.

the pressure inside the airway). Thus, elastic recoil of the lung is a major determinant of the cross-sectional area of the small airways.

The medium-sized bronchi contain smooth muscle in their walls. The tone of these muscles is determined by the balance of the sympathetic and parasympathetic stimulation they receive. Recall that sympathetic activity leads to muscle relaxation and dilation of the airway while parasympathetic simulation leads to muscle constriction and narrowing of the airway. Thus, bronchial smooth muscle tone is a major determinant of the cross-sectional area of the medium-sized airways.

QUICK CHECK 4-2   DETERMINANTS OF THE CROSS-SECTIONAL AREA OF THE AIRWAYS

Lung volume
Elastic recoil of the lung
Bronchial smooth muscle tone

Airway smooth muscle tension for a given level of autonomic nervous system activity varies through the respiratory cycle and is different after inhalation than exhalation, that is, it shows **hysteresis**. The airways are wider at a given lung volume when the volume is reached during deflation (i.e., coming down from a higher lung volume) than they are when the volume is reached during inflation (i.e., coming up from a lower lung volume). Conceptually, this phenomenon is similar to the hysteresis we described for the lung as whole in Chapter 3, but the mechanism that accounts for it is quite different. Airway hysteresis is believed to be the consequence of *stress relaxation* of the tissue. If you stretch tissue and hold it at that elongated position momentarily, the elastic forces relax just a bit. This effect appears to be accentuated when bronchomotor tone (smooth muscle tone) is increased, as during an asthma attack.

THOUGHT QUESTION 4-5: Imagine two patients who are suffering from an asthma attack. Initially, they have identical muscle tone in their medium-sized airways. For the next 10 minutes, you instruct the first patient to take shallow breaths to relatively low end-inspiratory lung volumes, and you tell the second patient to take large breaths to higher lung volumes. If you examine the cross-sectional area of the medium-sized airways at the end of the 10 minutes, what do you expect to find? Why?

TRANSMURAL PRESSURE AND THE SIZE OF AIRWAYS DURING FORCED EXHALATION

The transmural pressure of the airways—the pressure inside minus the pressure outside the airway (i.e., the pleural pressure)—is always positive during relaxed breathing at rest because pleural pressure is more negative than airway pressure throughout the respiratory cycle. Recall that a positive transmural pressure is called a distending pressure (this term is applicable any time the transmural pressure is positive, even as an airway decreases in diameter during exhalation; during exhalation, a positive transmural pressure across the airway means that the airway is not being compressed, even if pleural pressure becomes positive).

During conditions in which ventilation is elevated and expiratory flow must be increased, such as exercise, expiratory muscles may be recruited to push out air, and pleural pressure during expiration is often positive (i.e., greater than atmospheric pressure). If the pleural pressure is higher than the pressure inside the airway, transmural pressure becomes negative and the airway, if we are describing a small airway without smooth muscle or cartilage in its wall, may collapse. In the normal lung, whether or not the transmural pressure in the small airways becomes negative during exhalation is determined partly by how quickly the pressure in the airway declines as air travels from the alveolus to the mouth.

Exhalation begins when the inspiratory muscles relax and alveolar pressure becomes positive. Air begins to move out of the alveolus and flows toward the mouth. As the air moves down the airways, flow is constant. To sustain flow, pressure is dissipated because of the interplay of the factors we described earlier in this chapter as we examined the physical properties of flow through tubes.

First, pressure is lost as the resistance of the airways is overcome. The energy is dissipated as heat. Second, as you move from the periphery of the lung to the mouth, you go from millions of small airways in parallel ultimately to a single tube, the trachea. The total cross-sectional area of the airways diminishes. To maintain a constant flow, the velocity of the gas molecules must increase. In accord with Bernoulli's principle, this process leads to an additional decrease in pressure within the airway. Third, as you move from the peripheral to the central airways, you go from conditions of laminar flow to turbulent flow. To sustain a constant flow with this change in conditions, more work must be done; pressure decreases some more.

QUICK CHECK 4-3

FACTORS THAT LEAD TO A DECREASE IN AIRWAY PRESSURE DURING FORCED EXHALATION

Airway resistance
Bernoulli's principle
Transition from laminar to turbulent flow

Given that the pressure in the airways must decrease as air flows from the alveolus to the mouth, what determines whether or not transmural pressure becomes negative and the forces on the airway predispose it to narrow or collapse? To answer this question, let us consider the flow of water through an infinitely flexible tube, that is, a tube that will completely collapse as soon as the pressure outside the tube is greater than the pressure inside the tube (this type of tube is called a *Starling resistor*) (Fig. 4-10).

As you can see in Figure 4-10, water in container A is flowing through a tube and ultimately empties on the ground. The tube that exits the container, however, must pass through a second container of water, B. The water that exits container A has a pressure, P_I (the pressure inside the tube) that is initially proportional to the height of the water above the tube in container A. As the water flows through the tube, the pressure inside the tube, P_I, falls because of resistance in the tube. The point at which transmural pressure of the tube is zero is called the **equal pressure point** (EPP) because the pressure inside and outside the tube are now the same. If P_I becomes less than the pressure surrounding the tube (P_S) in container B, the tube collapses (transmural pressure is negative). When the tube collapses, flow stops. Use Animated Figure 4-10A to view this sequence up to the initial tube closure. What are the factors that determine the location of the point of collapse? What would happen if the animation were to continue? Think about these questions before reading on and finding out the answers.

In the absence of flow (with the tube collapsed), we are now back in static conditions, and the pressure in the tube at the point of collapse is now proportional to the height of the water in container A. P_I is now greater than P_S, the transmural pressure is positive, and

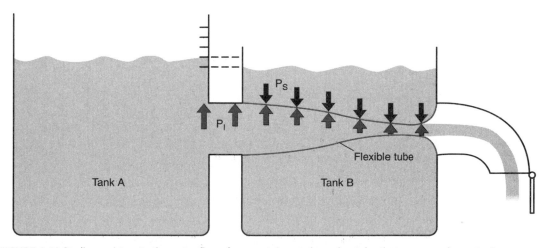

FIGURE 4-10 Starling resistor. As the water flows from container A through a tube that passes under water in container B, the pressure inside the tube, P_I, may become less than the pressure surrounding the tube, P_S. This negative transmural pressure leads to collapse of the tube and cessation of flow. See the text for more details. **Also see** Animated Figures 4-10A and 4-10B.

the tube opens and flow is reestablished. However, as soon as flow is reestablished, we are back in dynamic conditions and must deal with resistance in the tube. P_I again decreases, and collapse occurs. This cycle is repeated until sufficient water has exited the tube to make the height of the water in container A equal to the height in container B. Use Animated Figure 4-10B to view this sequence and observe the cycle of flow, pressure decrease, tube collapse, and static conditions leading to tube reopening and reestablishment of flow. You can use the slider to move slowly back and forth through the animation, as well as watch a detailed explanation by clicking on the ShowMe button.

Whether or not the P_S is greater than P_I before the tube exits container B depends on two factors, the differential in the height of the water in container A versus container B and the resistance of the tube. If we use this model as an analogy for the ventilatory pump, container A represents the alveolus, the tube represents the airways, and the pressure in container B, P_S, represents pleural pressure during a forced exhalation. For a given airway resistance and rate of decrease in pressure in the airway, the factor that determines whether transmural pressure (in this case P_I minus P_S) becomes negative in the airway is the difference between alveolar pressure and pleural pressure. As you recall from Chapter 3:

$$P_{alv} = P_{el} + P_{pl}$$

and

$$P_{alv} - P_{pl} = P_{el}$$

where P_{alv} = alveolar pressure; P_{el} = elastic recoil pressure; and P_{pl} = pleural pressure. Thus, the elastic recoil of the lung is a critical factor in determining whether airways collapse in the lung during exhalation.

Of course, the airways in the lungs are not Starling resistors, and the tendency of the small airways to collapse in the presence of a negative transmural pressure is offset, to some degree, by the tethering effect of the surrounding lung tissue, which is also a reflection of the elastic recoil of the lung. Furthermore, the pressure surrounding the alveolus and the airway in the thorax is the pleural pressure, which is essentially equal throughout the thorax (with minor differences based on gravity, as we discussed in Chapter 3). There is no true equivalent for the pleural pressure, with respect to its effect on alveolus and airway, within the Starling model. Nevertheless, the concepts represented by the Starling resistor are important in understanding the notion of *flow limitation* in the lungs, both in normal individuals and in those with diseases of the airways.

The Flow–Volume Loop and Flow Limitation

We have now reached the point at which we can integrate all of the isolated physical principles we have been describing with respect to flow through tubes and apply them to the specific circumstance of air movement in the lungs. As we develop the concepts of the flow–volume loop and flow limitation, you may find it useful to refer to the discussion of one or more of the principles outlined earlier in the chapter.

THE VOLUME–TIME PLOT

To measure the total amount of gas exhaled by a person, you ask her to breathe in until she cannot get any more into her lungs (she is now at total lung capacity [TLC]), and then exhale as hard and fast as possible until she cannot blow out any more (she is now at residual volume [RV]). The volume of the exhaled gas is the vital capacity (VC). To see how to plot the graph of volume versus time, look at Figure 4-11. Notice that the curve is steeper at the beginning (near zero exhaled volume) than near the end when the VC maneuver is complete. The slope of the curve at any point along the graph is equal to the flow at that point (flow = ΔVolume/ΔTime).

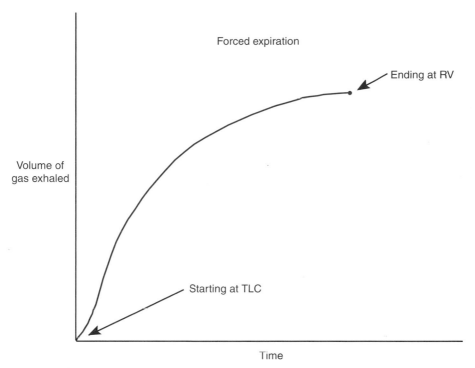

FIGURE 4-11 Volume–time plot of exhaled gas. This graph is a depiction of the amount of air exhaled from total lung capacity (TLC), shown at the base of the volume axis, to residual volume (RV). The subject has been instructed to exhale "as hard and fast as possible." Note that the slope of the graph, which is a representation of flow, is most steep near the beginning of the exhalation (when absolute lung volume is near TLC) and most flat at the end of the exhalation (when absolute lung volume is near RV).

Consider a new way of plotting this information. Take the flow, as determined from the slope of the volume–time plot at each point of the curve, and plot that flow as a function of the *absolute* volume at which it occurred (Fig. 4-12).

The patient is asked, at the end of a normal exhalation (functional residual capacity [FRC]), to take a deep breath in to TLC and then breathe out as hard and fast as possible until no more air will come out of the lung (i.e., RV). The resulting plot gives us an instant visualization of the inspiratory and expiratory flow at all lung volumes. Notice that the expiratory flow is greatest at the beginning of the expiratory maneuver and then gradually declines until RV is reached. The VC can be calculated as the distance between TLC and RV on the volume axis.

Two elements of this graph must be emphasized and understood to avoid confusion at a later time. First, each point along the outer line of the graph (i.e., the maximal forced maneuver) represents the maximal flow at a given volume. Second, time is not explicitly represented in this graph. Looking at the graph alone, you cannot determine whether the maneuver took 1 or 10 seconds. It is not possible, for example, to calculate from this graph how much air was exhaled during the first second of the expiratory maneuver (a quantity called the forced expiratory volume in 1 second [FEV_1]). Failure to remember these points will lead to errors in interpreting data from flow–volume loops in clinical practice.

Use Animated Figure 4-12A to see the flow–volume plot (*upper left*) being traced during normal inspiration and expiration. A spirogram is also shown (*center*), providing another view of the changes in lung volume translated onto a rotating roll of paper. Press the Show FVC button to instruct the subject to perform a forced vital capacity (FVC) maneuver and pay attention to the shape and time course of the resulting plots. Note how much

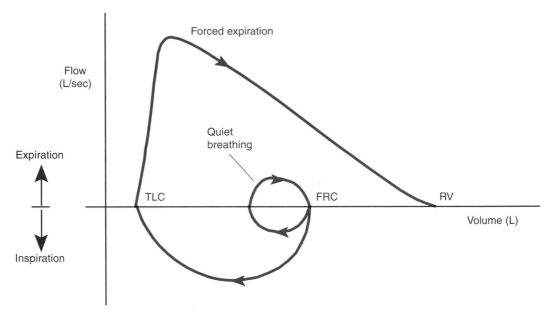

FIGURE 4-12 Flow–volume plot. This graph is a depiction of the flow of air into and out of the lungs as a function of the volume of gas in the lung. Notice the convention we use in depicting the relationship between flow and volume. The total lung capacity (TLC) is depicted to the left and the residual volume (RV) to the right on the volume axis. Expired flow is "up" on the graph, and inspired flow is "down." The small loop in the center of the graph represents normal, resting breathing. The subject is then instructed to take a "deep breath in until no more air can be inhaled, and then blow out as hard and fast as possible." Notice that the expired flow is greatest at the beginning of the exhalation and then gradually diminishes until RV is reached. **Also see Animated Figure 4-12A.**

air is exhaled in the first second of the forced maneuver. There is a great deal going on in this diagram, and you may find it helpful to use the Pause button to freeze the motion at certain points.

Animated Figure 4-12B demonstrates that the expiratory volume–time plot and flow–volume plot are simply different viewpoints on a forced expiration. Upon initial viewing of the three-dimensional diagram, the first graph that appears is a flow–volume plot (shown on the left as a shaded surface and on the right as a line graph, analogous to the upper half of Figure 4-12). Note that the maximal flows occur at high lung volumes but also observe that this perspective has no time axis, so we cannot tell what fraction of air was exhaled in the first second. Now drag the horizontal slider all the way to the left, and you will see the flow–time plot come into view. Again, note how flow is maximal very near the beginning of the forced expiration and then declines slowly over the course of the expiration.

Now drag the vertical slider all the way to the top, and you will see the volume–time plot come into view (analogous to Figure 4-11). Observe that the steepest slope is near the beginning of expiration, and this corresponds to the time when flow is maximal (the flow is equal to the slope of the volume-versus-time graph).

MAXIMAL EXPIRATORY FLOW: THE FORCES TO BE CONSIDERED

As discussed earlier in this chapter, flow through a tube is determined by the driving pressure (i.e., the difference in pressure between the two ends of the tube) and the resistance of the tube.

$$\dot{V} = \Delta P / R$$

When considering the driving pressure during a forced exhalation, we must examine the factors that determine alveolar pressure (one end of the tube). The other end of the tube is the mouth, where the pressure is always equal to atmospheric pressure, or 0 by convention.

The alveolar pressure is the result of the elastic recoil of the lung and the pleural pressure.

$$P_{alv} = P_{el} + P_{pl}$$

Elastic recoil of the lung is determined by the intrinsic properties of the lung and, for a given lung, by the volume of the lung. The greater the lung tissue is stretched, the greater the elastic recoil forces. Surface forces, related partly to the density of surfactant in the surface layer, as discussed in Chapter 3, also contribute to the recoil of the alveoli. The density of surfactant in the surface layer tends to be less at TLC than at lower lung volumes during exhalation. Surface forces, therefore, are greatest at TLC. The pleural pressure during exhalation is determined by the position of the chest wall relative to its relaxation volume and the stiffness or recoil of the chest wall. In addition, you must consider the forces that can be generated by the expiratory muscles. Pleural pressure is greatest (most positive) when a person attempts to do a forced exhalation from TLC. At this volume, the chest wall is farthest above its relaxation volume, and the expiratory muscles are at their greatest length (hence, in the best position to generate tension). Consequently, alveolar pressure, and therefore, driving pressure, is greatest at TLC.

The resistance of the airways varies with their diameter. The cross-sectional area of the airways, particularly the small airways, varies with the size of the lung. The greater the lung volume, the greater is the diameter of the airways. Thus, airway resistance is least when the lung volume is at TLC. Taken together, the factors that determine the driving pressure and the resistance of the lung explain why expiratory flow is maximal at TLC.

QUICK CHECK 4-4   **FACTORS THAT DETERMINE MAXIMAL EXPIRATORY FLOW**

Elastic recoil of the lung and chest wall
Surface forces in the alveoli
Force of contraction of the expiratory muscles
Resistance

Note: all of these factors are affected by lung volume.

EXPIRATORY FLOW LIMITATION

As you view Figure 4-12, you notice that the flow diminishes as lung volume decreases. Part of this decrease in flow is because of the fact that the driving pressure decreases as lung volume declines; the elastic recoil of the lungs and chest wall is less at lower lung volumes, and the expiratory muscles are shorter, thereby reducing their potential to generate tension. Furthermore, the airway resistance increases as lung volume declines because the size of the small airways depends on the pull of the surrounding lung tissue. But what if you could strengthen your expiratory muscles? Could you generate a flow at FRC that would equal the flow at TLC? To answer this question, we must take a closer look at what happens to the transmural pressure across the airways during a forced exhalation.

Imagine you are at a lung volume between FRC and TLC. You begin a forced exhalation. The expiratory muscles contract, squeezing the pleural space and raising pleural pressure, thereby generating a maximal pressure in the alveolus. A pressure differential between the alveolus and the mouth now exists, and air exits the alveolus to begin its journey to the mouth. The gas exits the alveolus at a flow determined by the driving pressure and the resistance of the airway; assume that this initial flow is constant until the gas exits the mouth. As the gas moves along the bronchial tree toward the mouth, a loss of pressure occurs because of airway resistance, the increasing velocity (remember the Bernoulli effect and the analogy of the multiple one-lane highways merging into a single two-lane highway), and the transition from laminar flow in the peripheral airways to the turbulent flow of the central airways (see Quick Check 4-3).

As pressure within the airway declines, you may eventually arrive at a point at which the pressure inside the airway is the same as the pressure outside the airway (pressure outside the airway is essentially equal to pleural pressure); at that point, transmural pressure is 0. Any further loss of pressure within the airway leads to a negative transmural pressure and the compression or collapse of the airway. The point at which transmural pressure of the airway is 0 is called the EPP (recall the Starling resistor) because the pressure inside and outside the airway are now the same. If the EPP occurs in the most central airways of the lung, as is the case in normal, healthy lungs, where the walls of the airway are largely supported by cartilage, the effects of a negative transmural pressure are minimized. In contrast, if the EPP occurs in the small, peripheral airways, as may be the case in diseased lungs (e.g., in emphysema), the development of a negative transmural pressure leads to collapse of the airway (Fig. 4-13).

If the EPP occurs in a relatively unsupported airway and collapse occurs, flow transiently stops. As we described in our discussion of the Starling resistor, when flow ceases, static conditions are established in the airway, and the pressure equalizes between the alveolus and the EPP. Transmural pressure is now positive again, and the airway opens.

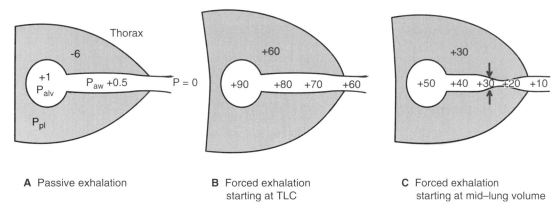

A Passive exhalation   **B** Forced exhalation starting at TLC   **C** Forced exhalation starting at mid–lung volume

FIGURE 4-13 Flow limitation and the equal pressure point. (All pressures shown are in cm H_2O). **A,** Here we see the pressures in the pleural space (P_{pl}), alveolus (P_{alv}), and airway (P_{aw}) during a passive exhalation. Note that the P_{pl} is negative throughout the exhalation and the transmural pressure of the airway is always positive; there is no compression of the airway. **B,** A forced exhalation beginning at total lung capacity (TLC). In this case, the point at which the pressure outside the airway (pleural pressure) is the same as the pressure within the airway (the equal pressure point [EPP]) occurs in the most central airways of the lung (where pressure is +60 in the diagram). Collapse does not occur because these airways are well supported by cartilage. **C,** A forced exhalation beginning at a lung volume between functional residual capacity (FRC) and TLC. The EPP develops in a more peripheral (collapsible) airway (where pressure is +30 in the diagram), the airway is compressed, and flow limitation is established. One recent study suggests that even when starting at TLC, flow limitation is present with a maximal expiratory maneuver. **Also see Animated Figure 4-13.**

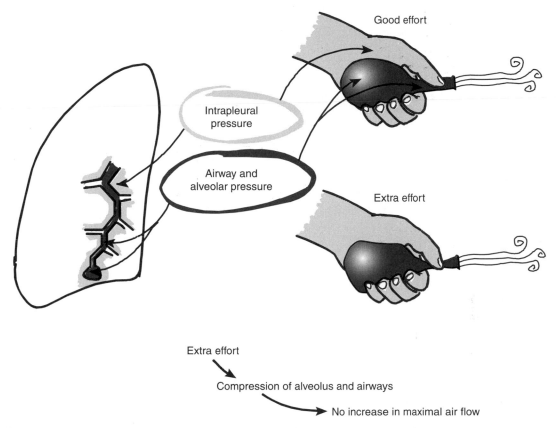

FIGURE 4-14 Flow limitation. The hand squeezing a balloon is shown here as an analogy for expiratory effort during a forced expiration. Extra effort leads to increased pleural pressure, but no increase in maximal air flow because the pleural pressure acts on both the alveolus (shown as the hand pressing on the balloon) and the airway (shown as the thumb pressing on the neck of the balloon).

Flow is reestablished, and pressure in the airway decreases, which again leads to the development of the EPP.

As soon as the EPP has occurred, the driving pressure that determines flow is no longer the difference in pressure between the alveolus and the mouth. Rather, the driving pressure is now the difference between alveolar pressure and pleural pressure as the airway cycles between open and closed states. Trying to exhale more forcefully by increasing the force of contraction of the muscles, and thereby the pleural pressure, has no effect. If pleural pressure increases, alveolar pressure, which is the sum of elastic recoil pressure and pleural pressure, must increase by the same amount; the driving pressure remains the same (Fig. 4-14). Under these conditions, one has achieved **flow limitation**, that is, maneuvers to produce an increase in pleural pressure do not yield an increase in flow. When flow limitation is present, the driving pressure is equal to the elastic recoil of the alveolus.

Whether EPP develops in a peripheral or central airway depends on the difference between the alveolar pressure and the pleural pressure at the beginning of the expiratory maneuver. The difference between alveolar and pleural pressure is the elastic recoil pressure. Thus, elastic recoil pressure is the determining factor for the location of the EPP. To the extent that the exhalation is at a lower lung volume, EPP is more likely to be in the peripheral airways. In disease states such as emphysema, in which the elastic recoil of the lungs is diminished, the likelihood of reaching EPP in the peripheral airways is increased for any lung volume (Fig. 4-15).

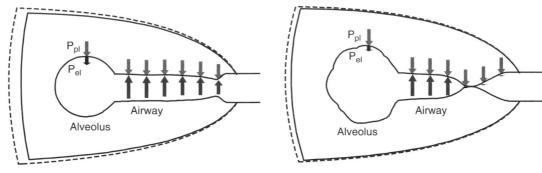

A Normal forced expiration from TLC   **B** Emphysema forced expiration from TLC

FIGURE 4-15 Location of the equal pressure point (EPP). The location of the EPP depends on the difference between the alveolar pressure and the pleural pressure (i.e., the elastic recoil pressure) near the beginning of the expiratory maneuver. **A,** In a normal forced expiration, elastic recoil starts out high, and the EPP occurs in more central airways. **B,** In disease states such as emphysema, in which the elastic recoil of the lungs is diminished, the EPP occurs in more peripheral, collapsible airways. TLC = total lung capacity; P_{pl} = pleural pressure; P_{el} = elastic recoil pressure. **Also see Animated Figure 4-15.**

Use Animated Figure 4-15 to observe the effect of elastic recoil on the location of the EPP. Before playing the animations, compare emphysema with the normal state by using the selection buttons below the slider and observe the difference in the elastic recoil pressures (magnitude of the *arrow* shown in alveolus). Use the play button or slider to watch the animations and notice how the lower elastic recoil pressure for emphysema leads to an EPP in more peripheral airways. Remember that the animations are of *forced* expirations, so pleural pressure starts out negative (for the initial static state, with negative pressure shown as the *arrow* pointing away from the alveolus) and rapidly becomes positive as the expiratory muscles apply force.

PRESENCE OF FLOW LIMITATION ON THE FLOW–VOLUME LOOP

The flow–volume loop is one of the essential elements of the tests of clinical pulmonary function. A firm understanding of the physiology underlying this test is critical for assessing a range of lung diseases.

Developing the Flow–Volume Loop

To begin, take a deep breath in until you have reached TLC and then exhale as hard and fast as you can until you are at RV. Collect and measure the volume of exhaled gas and plot the volume as a function of time (see Fig. 4-11). The slope of the curve, which represents flow ($\Delta V/\Delta t$), is most steep at the beginning of the exhalation, when you are exhaling near TLC. In contrast, as you near RV, the slope gradually diminishes.

As we described previously, by plotting the instantaneous slope as a function of absolute volume, you can generate a graph that represents maximal flows for all volumes during the expiratory maneuver. Recall that the amount of gas exhaled during the forced expiratory maneuver (from TLC to RV) is the forced vital capacity (FVC). (Fig. 4-16, also see Animated Figure 4-12A)

Once you have reached RV, if you inspire as hard and fast as you can back to TLC and again plot the flow as a function of lung volume, you have created the inspiratory portion of the curve. If you now relax and breathe normally, you will see a smaller curve within the envelope of the maximal inspiratory and expiratory maneuvers (Fig. 4-17).

Each point on the outer curve represents the maximal flow achievable at a given lung volume. Recall that time is not represented on this graph. You cannot calculate the FEV_1 from this graph. We are visualizing only flow at different lung volumes.

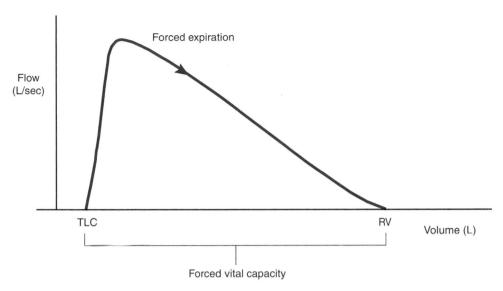

FIGURE 4-16 Expiratory flow–volume plot. The graph represents the instantaneous slope, or flow, taken from each point of the volume–time curve and plotted as a function of the absolute lung volume. Note that by convention, we place the total lung capacity (TLC) toward the lefthand portion of the volume axis and the residual volume (RV) toward the righthand portion.

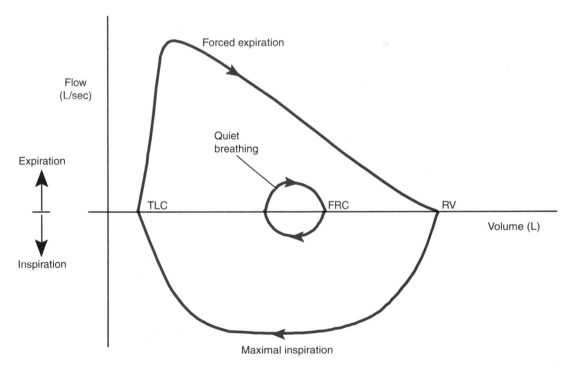

FIGURE 4-17 Full flow–volume plot. This graph represents the full maximal exhalation (from total lung capacity [TLC]) and inhalation (from residual volume [RV]) demonstrated by the outer curve, as well as resting breathing demonstrated by the inner curve. Note that by convention, expiratory flow is "positive" on the flow axis, and inspiratory flow is represented as "negative." The volume at the end of exhalation during resting breathing is functional residual capacity (FRC).

As you examine the normal flow–volume loop, observe the following. First, the maximal expiratory flow is somewhat greater than the maximal inspiratory flow. Second, the maximal inspiratory flow occurs approximately halfway between RV and TLC, but the maximal expiratory flow is very close to TLC. Third, the decline in flow from the point of maximal flow until the end of exhalation is relatively linear. To understand these findings, we need to reconsider the forces (i.e., muscle contraction, recoil of the lung, and recoil of the chest wall) that generate a pressure differential across the airways and the factors that impede flow (i.e., resistance, size of airways, turbulence). We will use the principles we have outlined in the early sections of this chapter.

Expiratory flow is maximal near TLC when recoil inward of the lungs and chest wall is greatest, when the muscles of exhalation are longest and thus best able to generate tension, and when the airways are at their greatest diameter. Maximal inspiratory flow is achieved at a lung volume that represents a tradeoff between the length of the inspiratory muscles (longest at RV), the recoil of the chest wall outward (greatest at RV) and the lung inward (least at RV), and the diameter of the airways (smallest at RV). The decline in the maximal expiratory flow with decreasing lung volume, known as expiratory flow limitation, reflects the balance of airway and pleural pressures, that is, the transmural pressure across the airways.

Expiratory Flow Limitation: Passive Versus Forced Exhalation

We will now reexamine the concept of flow limitation and use it to understand the shape of the flow–volume loop. Some of this may seem repetitive, but these are important concepts for the interpretation of pulmonary function tests and for a full appreciation of normal physiology as well as the pathophysiology of obstructive lung disease.

First, let us examine a passive exhalation from a lung volume approximately 0.5 L above FRC, that is, at the end of a normal resting inhalation (see part A of Animated Figure 4-13). At the start of the expiration, alveolar pressure is 0 because we are under static conditions (no flow) with the airway open. As previously described, during the expiration, alveolar pressure is positive because of the recoil of the lung (secondary to elastic and surface forces), and pleural pressure is slightly negative because of the opposing forces of the lung and chest wall. The difference in pressure between the alveolus and the mouth propels air into the airway. As air moves along the airway from the alveolus to the mouth, the gas has to overcome resistance, leading to a loss of pressure in the airway. In addition, as air proceeds from millions of small peripheral airways to a lesser number of more central airways, the velocity of the gas must increase for flow to be constant. In accord with Bernoulli's principle, this results in a further loss of pressure within the airway. Finally, as you transition from laminar to turbulent flow in the central airways, an even greater pressure differential (in other words, pressure decrease), is necessary to maintain a constant flow. Thus, as air travels from the alveolus to the mouth during the expiratory maneuver, the intra-airway pressure progressively diminishes. The pleural pressure, which surrounds the alveolus and the airways, is negative throughout the entire passive exhalation near FRC because the chest wall is below its resting volume and is recoiling outward. As a result, there is always a positive transmural pressure (pressure inside greater than pressure outside) across the airway despite the loss of pressure within the airways. Flow is unencumbered by collapse of airways. Use part A of Animated Figure 4-13 to view a passive expiration, and pay special attention to the transmural pressure (difference between pressure inside the alveolus or airway and pressure outside, or pleural pressure). Think about the following questions before you continue reading the text: What would be different about pleural pressure during a forced expiration? How might this affect transmural pressure?

Now contrast the passive exhalation with what occurs during a forced expiratory maneuver that begins near TLC (see part B of Animated Figure 4-13). The pleural pressure at the very beginning of the maneuver (static state) is negative, a consequence of the collapsing force on the lung and the net expanding force on the chest wall at this volume (inspiratory muscles pulling out on the chest wall; elastic recoil of chest wall inward at TLC). Taken together, these forces tend to pull apart the pleural surfaces. The recoil pressure of the lungs is approximately 30 cm H_2O at TLC.

The expiratory muscles are now activated. Assume that the pleural pressure during the forced expiration reaches 60 cm H_2O (note: pleural pressure is now positive because the chest wall is above its resting volume and is recoiling inward and because you are activating expiratory muscles to squeeze the chest wall, pleura, and lungs). This means that the pressure within the alveolus (recall: $P_{alv} = P_{el} + P_{pl}$) reaches 90 cm H_2O. As before, we assume a constant flow must be maintained from the alveolus to the mouth.

The pressure around the airway, as already noted, is essentially equivalent to the pleural pressure. For the purpose of this analysis, the pleural pressure may be viewed as constant throughout the thorax. The pressure in the alveolus is greater than the pressure in the pleural space ($P_{alv} = P_{el} + P_{pl}$), but as the pressure in the airway diminishes with the flow of gas from alveolus to mouth, one may now reach a point at which the pressure inside and outside the airway are the same (in our example, the airway pressure has decreased from 90 to 60 cm H_2O). This point, known as the EPP, is the point at which transmural pressure across the airway is 0. Use part B of Animated Figure 4-13 to view the change in pressures during a forced expiration starting from TLC (only shown up through development of the initial EPP), and pay special attention to the location where the airway pressure first becomes equal to pleural pressure. If the airway were an infinitely flexible tube, as we examined with the Starling resistor, the airway would now collapse, and flow would cease. Pressure in the airway would then equalize with the alveolus (static conditions in place), the transmural pressure would become positive, the airway would reopen, and flow would be reinstituted. As soon as flow of gas was restarted, however, the EPP would reappear, and the cycle would repeat.

As soon as the EPP has been established, further increases in pleural pressure (i.e., blowing more forcefully) does not lead to an increase in expiratory flow. Although the increase in pleural pressure will increase the alveolar pressure and thus, the intra-airway pressure, it also increases the pressure surrounding the airway (see Fig. 4-14). Consequently, transmural pressure is not altered. To the extent that further increases in pleural pressure do not yield greater expiratory flow, these conditions are sometimes termed **effort independence**. At the EPP, the driving pressure is no longer the alveolar pressure minus the pressure at the airway opening. As noted in the Starling resistor analogy, the driving pressure is now the alveolar pressure minus the pleural pressure.

The concept of the EPP depends on conceptualizing the airways as highly flexible, similar to a Starling resistor. However, we know from Chapter 2 that the central airways of the lung, especially the trachea, are supported by cartilaginous rings that help resist deformation when transmural pressure is negative (i.e., pressure outside greater than pressure inside the airway). What determines whether the EPP occurs within a flexible peripheral airway or a supported central airway?

The key factor in determining where along the airway the EPP will occur for a given pulmonary resistance and pulmonary compliance is the difference between the alveolar pressure and the pleural pressure. When the difference between these pressures is large, gas will travel a long distance toward the mouth before losing sufficient pressure to achieve EPP. Alternatively, if the alveolar pressure and the pleural pressure are nearly the same, a small decrease in pressure will result in EPP. The difference between alveolar pressure and pleural pressure is the elastic recoil pressure. For a given lung, elastic recoil pressure changes in concert with the volume of the lung; at higher lung volumes, elastic recoil pressure is greater. Therefore, EPP tends to occur toward the supported, central airways when gas is exiting the alveolus at high lung volumes and in the peripheral, unsupported airways when gas exits the alveolus at low lung volumes. Take another look at Animated Figure 4-13, this time paying special attention to the lung elastic recoil pressure (shown above each lung). Note how the lower elastic recoil pressure in part C leads to an EPP in the more peripheral, unsupported airways.

Flow Limitation and the Flow–Volume loop

Given what you have just learned, reexamine Figure 4-17. The vast majority of the expiratory flow–volume loop (the portion to the right of the peak flow) is constrained by the principles of flow limitation. What does this mean? Pick a point on the maximal expiratory flow curve to the right of the peak flow. At that flow and that lung volume, further increases in expiratory effort (or pleural pressure) will not result in a higher flow. If you pick a point near RV, the flow at which flow limitation is present is very low, much lower than at a volume near FRC. Stated differently, the same expiratory effort at FRC may yield a greater flow than at RV. As discussed, this is largely because of the differences in the elastic recoil pressure at these two lung volumes. New experimental data suggest that even at lung volumes near TLC, we may be flow limited as well.

THOUGHT QUESTION 4-6: How would you construct an experiment to determine whether normal individuals are flow limited even at TLC? In other words, how would you determine whether greater effort would or would not lead to greater flow near TLC?

THOUGHT QUESTION 4-7: How would the flow–volume loop be affected by a disease process that reduced elastic recoil of the lung?

Emphysema and the Flow–Volume loop

In emphysema, lung tissue is destroyed and the elastic recoil of the lung is reduced. This process has several effects on the factors that determine expiratory flow. As you may recall, the small, flexible, peripheral airways are supported by the recoil of the lung tissue through which they pass (review Figure 4-9). This support helps to prevent narrowing of the airways when transmural pressure is negative, as occurs during a forced exhalation. With the weakening of these supports, the small airways are more likely to collapse under these circumstances.

Elastic recoil pressure is one of the major determinants of alveolar pressure. Therefore, a reduction in elastic recoil reduces the driving pressure during exhalation. Furthermore, the location of the EPP (i.e., whether it will occur toward the more central airways or be found in the small, peripheral airways) is largely determined by the difference between the alveolar pressure and the pleural pressure. A smaller difference, which results from a lower elastic recoil pressure, predisposes to the movement of the EPP toward the alveolus regardless of the lung volume. Both of these factors contribute to the development of **dynamic compression** (reversible narrowing of airways that results from changes in transmural pressures during exhalation) of the airways at higher lung volumes than is seen in normal individuals. Consequently, in a patient with emphysema, flow limitation may exist at any given lung volume between TLC and RV.

The flow–volume loop also takes a different shape in patients with emphysema. The initial peak flow is lower than normal because of the reduced driving pressure seen in the

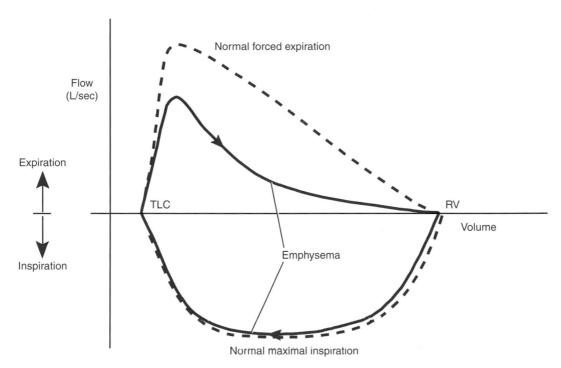

FIGURE 4-18 Flow–volume loop in emphysema. Note that expiratory flow is reduced at all lung volumes. This may occur either because of increased narrowing of airways or because of the reduced driving pressure in the emphysematous lung; both of these may occur because of the reduced elastic recoil of the lung tissue in emphysema. Remember that the elastic recoil pressure of the alveolus is a key determinant of the driving pressure during exhalation and of the difference between alveolar and pleural pressure (the difference between P_{alv} and P_{pl} plays a major role in the location of the equal pressure point). (Note: the TLC and RV may be altered in emphysema [typically, both are increased due to loss of elastic recoil in the lungs], but are shown normalized to the same range as the normal lung in order to better compare the flows and shapes of the curves.) RV = residual volume; TLC = total lung capacity.

alveolus at TLC and because, even at TLC, there may be dynamic compression of the airways. In addition, the expiratory flows decrease rapidly at lung volumes below TLC because of further narrowing of the airways. As a result, the expiratory curve, rather than appearing fairly linear from the peak flow to RV, takes on a concave-upward appearance, termed **expiratory coving** (Fig. 4-18).

The inspiratory portion of the curve is relatively normal in appearance. During inspiration, transmural pressure across the airways is positive. Therefore, the loss of elastic recoil does not disturb inspiratory flow.

THOUGHT QUESTION 4-8: Examine the flow–volume loop of the patient with emphysema shown in Figure 4-18. Describe what must happen to the way that a patient with emphysema breathes in order to exercise.

TIME CONSTANTS

A **time constant** is a measure of how well gas is moved in and out of alveoli. A large time constant implies that it takes a long time to exchange gas between the atmosphere and the alveolus; a short time constant implies a rapid exchange. The time constant can be conceptualized as the product of the airway resistance and the lung compliance.

$$\text{Time constant} = \text{Resistance} \times \text{Compliance}$$

An area of lung in which the airway resistance is high will have a large time constant because it will take longer for air to move into and out of that region of the lung for any given driving pressure (i.e., flow will be reduced). An area of lung with a high compliance (i.e., with reduced elastic recoil) will have a large time constant because the elastic recoil of the lung is a primary determinant of the driving pressure during exhalation; expiratory flow will be reduced.

Let us consider a clinical example to demonstrate the implications of this concept. You ask a person to breathe 100% oxygen. What happens to the mixture of gases in the alveolus? Just before the first breath of oxygen, the alveolus contains mostly nitrogen, some oxygen, and a small amount of carbon dioxide. The amount of gas in the lungs at FRC may be 3 to 4 L, depending on the size and age of the person. The individual now inhales a tidal volume (approximately 0.5 L) of 100% oxygen. Even if the lungs are normal, the gas in the alveolus will not immediately reflect the change in the concentration of gases in the inspired air. It will take several minutes before the nitrogen has been washed out of the lungs (recall the nitrogen washout technique for measuring FRC in Chapter 3). If the time constant of the lungs is very high, it may take 15 to 20 minutes before the gas in the alveoli reflects the change in the concentrations of the inspired gases.

The notion of time constants is used to conceptualize certain properties of the lung (you will never be asked to calculate the time constant in a clinical setting). Because most diseases of the lung are heterogeneous in their distribution, we tend to speak of the time constant for a region of lung rather than the lung as a whole. For example, a patient with emphysema has some areas of the lung that are relatively normal and other areas that are quite abnormal because of the disease process.

The concept of time constants offers us another physiologic explanation for the shape of the flow–volume loop in a patient with emphysema (see Fig. 4-18). Assuming a heterogeneous distribution of time constants throughout the lung, the units with a relatively normal time constant will have high expiratory flows and will empty first during the forced exhalation (the high flows near TLC). Next, the units that are only mildly affected will empty with intermediate flows. Finally, the most diseased units will empty very slowly with low flows (the flows seen near RV).

PUTTING IT TOGETHER

A 55-year-old man comes to see you as an outpatient for evaluation of shortness of breath. The patient indicates that he mostly has a problem when he exercises, but he has also noted some episodes of breathlessness at rest, particularly in the spring and fall. His family members have told him that they hear him wheezing at times, although he is less aware of that. Sometimes his shortness of breath feels like a sensation of "tightness" in the chest. At other times, he has a sensation of increased effort to breathe and a sense of not being able to get a deep breath. The patient has smoked two packs of cigarettes a day for the past 25 years. When you examine his chest, you note that his breath sounds are diminished and that it takes longer for him to exhale than inhale. You hear faint expiratory wheezes. Pulmonary function tests are performed and a flow–volume loop is obtained. The patient is given a dose of a β-agonist inhaler, and the flow–volume loop is repeated (Fig. 4-19).

The basic shape of the loop is indicative of emphysema with a concave-upward pattern. However, the improvement in the expiratory flow at each lung volume after using the inhaled bronchodilator suggests that there is also an element of reversible airway narrowing caused by bronchospasm. This airway narrowing is attributable to a component of asthma and probably explains the symptoms the patient experiences at rest in the spring and fall when allergens are

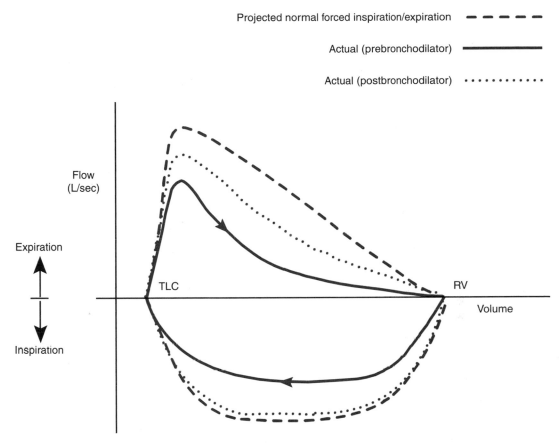

Projected normal forced inspiration/expiration   – – – – –

Actual (prebronchodilator)   ——————

Actual (postbronchodilator)   ············

FIGURE 4-19 Patient's flow–volume loop before and after bronchodilator use. The initial loop shows reduced expiratory and inspiratory flow at all lung volumes. In addition, the expiratory loop has a concave-upward shape (expiratory coving) indicative of emphysema. After use of the bronchodilator, the inspiratory flow is normal. The expiratory flow is improved but still less than normal at all lung volumes. (Note: as for Figure 4-18, the TLC and RV may be altered in emphysema [typically, both are increased due to loss of elastic recoil in the lungs], but are shown normalized to the same range as the normal lung in order to better compare the flows and shapes of the curves.) RV = residual volume; TLC = total lung capacity.

prevalent in the environment. The diminished breath sounds on the physical examination are a sign of a reduced number of airways in the lung due to the destructive nature of the emphysematous process. The prolonged expiratory phase of the respiratory cycle is secondary to expiratory airflow obstruction attributable to dynamic compression of the airways. The patient has a sense of difficulty getting a deep breath with exercise because he must hyperinflate his chest to increase ventilation with exercise, and his inspiratory capacity becomes limited by TLC—he literally cannot take a deep breath. You start the patient on regular bronchodilators and advise him to stop smoking.

Summary Points

- Dynamics is the study of the respiratory system during conditions of flow of gas into and out of the lungs.
- Driving pressure is equal to flow multiplied by resistance.
- Under conditions of laminar flow, changes in driving pressure are proportional to changes in flow.
- Under conditions of turbulent flow, changes in driving pressure are proportional to the square of the flow.

- The Reynolds number predicts whether flow in tubes is laminar or turbulent. Gases of lower density are more likely to be associated with laminar flow, as are gases traveling at low velocity.
- Poiseuille's law states that airway resistance is inversely proportional to the fourth power of the radius. Small increases in the radius of an airway greatly diminish the resistance in that airway.
- The arrangement of tubes that are connected is important in determining the total resistance of the system. For tubes arranged in series, the total resistance is the sum of each of the individual resistances. For tubes arranged in parallel, the total resistance is less than the resistance of any of the individual tubes. Because of the parallel arrangement of the millions of small airways in the periphery of the lung, most of the lung's resistance is located in the more central airways.
- As air moves from the millions of small airways in the periphery of the lung to the single trachea, the velocity of the gas increases. This phenomenon contributes to the presence of turbulent flow in the central airways.
- Bernoulli's principle states that as the velocity of a gas increases, the pressure exerted by the gas decreases. This phenomenon contributes to the decrease in the pressure in the airway during exhalation as air moves from the alveolus to the mouth.
- Small airways in the periphery of the lung are tethered open by the elastic recoil of the lung tissue through which they travel.
- The smooth muscle in the walls of medium-sized airways is under the control of the autonomic nervous system. Bronchial smooth muscle tone is a major determinant of the cross-sectional area and, consequently, the resistance of the medium-sized airways.
- The transmural pressure across the airway is always positive in a normal person during quiet breathing. During a forced exhalation, transmural pressure may become negative, which predisposes to airway narrowing and collapse in small, noncartilaginous airways.
- During a forced exhalation, pressure in the airway decreases as gas moves from the alveolus to the mouth because of several factors, including airway resistance, Bernoulli's principle, and the development of turbulent flow. When the pressure in the airway equals the pleural pressure, the transmural pressure is 0, the EPP is reached, and airway compression may follow.
- If the EPP is reached in peripheral airways, the condition of flow limitation exists; further increases in pleural pressure will not result in an increased flow.
- The elastic recoil of the lung is a major determinant of the location of the equal pressure point. Low elastic recoil pressure, caused by low lung volumes or diseases such as emphysema, shifts the equal pressure point toward the periphery of the lung.
- Most of the expiratory portion of the flow–volume loop represents conditions of flow limitation; at a given volume, as soon as the flow associated with a maximal expiratory effort at that volume is reached, further increases in pleural pressure will not result in increased flow.
- Time constants tell us about the rate at which gas is exchanged between an alveolus and the atmosphere.

Answers TO THOUGHT QUESTIONS

4-1. Turbulent flow leads to vibration of the airways as the molecules of gas bounce against the walls. These vibrations produce sounds that we call wheezes. In patients with asthma or emphysema who have narrowing of the airways, the velocity of the gas increases as it passes through these airways, and wheezing can be heard.

4-2. Laryngospasm can be a life-threatening medical problem. Treatment with bronchodilators often takes hours to work. In the meantime, you can look at parameters other than radius that affect Reynolds number and devise an intervention to buy time. By administering a gas with a lower density than air, one can transform turbulent flow to laminar flow. We achieve this by asking the patient to breathe a mixture of helium and oxygen (called heliox) that has a lower density than the nitrogen–oxygen mixture that constitutes the atmosphere. They may sound like Donald Duck when they talk, but they feel better.

4-3. Most people breathe through their noses during quiet breathing. The nose serves an important role in the humidification of gas that enters the lungs and helps to filter out particles in the gas (as anyone with seasonal allergies knows when the pollen count increases). However, the diameter of the passages through the nose is relatively small, so the nose imposes a substantial resistance to flow. During aerobic exercise, as ventilation increases, we open our mouth and breathe in a way that bypasses the nose to reduce resistance and the work of breathing.

4-4. The pressures under the two conditions are the same at the beginning and end of the inflation because, at these points, there is no flow, and the relationship between pressure and flow is reflecting only the compliance of the lung, which is not affected by the manner used to get from V1 to V2. However, achieving the change in volume under conditions of high flow requires more work because you must overcome airway and tissue resistance, factors that are not relevant in the static state.

4-5. The second patient will have larger airways at the end of the 10-minute trial. By inhaling to larger tidal volumes, there will be a greater stress-relaxation effect on the airways, and the muscle tone of the airways will be reduced. This effect can be seen in measurements of lung function, especially expiratory flows, in individuals with asthma.

4-6. If flow limitation is not present, it implies that the generation of a greater pleural pressure, and thus, a greater alveolar pressure and alveolar to mouth pressure gradient, will result in an increase in flow. But if someone is already doing a maximal effort at TLC, how do you create an even greater pressure gradient between the alveolus and the mouth? One approach to this has been to ask subjects to do a maximal expiratory maneuver from TLC while breathing on a mouthpiece that is attached to a device that can create a negative pressure at the mouth. Thus, by decreasing pressure at the airway opening, one can increase the pressure gradient from alveolus to the mouth during a maximal expiratory maneuver. When this technique is used, subjects appear to be flow limited, even at TLC (this technique and these results are somewhat controversial and need to be replicated before all will agree that flow limitation exists at TLC).

4-7. With reduced elastic recoil, the alveolar pressure at any given lung volume will be reduced compared with normal conditions. This will reduce the peak flow at the beginning of exhalation as well as all flows during the exhalation. Furthermore, because the smaller airways are tethered open by the elastic recoil of the surrounding lung tissue, loss of elastic recoil causes them to be more likely to narrow when pleural pressures are high, further reducing flow as lung volume gets smaller during the exhalation. This effect leads to a change in shape of the expiratory portion of the loop—it becomes concave upward, referred to as "expiratory coving"; (see Fig. 4-18).

4-8. When you exercise, you must increase ventilation. To increase ventilation, you must increase flow. But if you have severe emphysema, then even during resting breathing, you may already be operating at maximal expiratory flow over that range of lung volumes. (Graphically, we see this by observing from the flow–volume plot that for severe emphysema, the maximal expiratory flow–volume loop overlaps the quiet breathing loop.) Therefore, if you have severe emphysema, the only way to increase flow (necessary for exercise) is to increase the volume at which you are breathing (i.e., to hyperinflate the lungs—look again at the loop). The work of breathing is increased at higher lung volumes, partly because the muscles of inspiration are shorter (remember the length–tension relationship for muscles) and partly because one is operating on a stiffer portion of the pressure–volume curve of the respiratory system. But those with severe emphysema must hyperinflate if they are going to exercise under these conditions.

Review Questions

DIRECTIONS: *Each of the numbered items or incomplete statements in this section is followed by answers or by completions of the statement. Select the ONE lettered answer or completion that is BEST in each case.*

1.   You are taking care of a patient in the emergency department. She was placed on oxygen because of a complaint of shortness of breath. The patient appears to have fairly severe emphysema. You would like to do an arterial blood gas, a blood test that measures the amount of oxygen in the blood, but you are interested in what the oxygen level is when the patient is breathing regular air without supplemental oxygen. You take the supplemental oxygen away from the patient. How long should you wait after the oxygen has been removed until you obtain the arterial blood gas to ensure that you get an accurate measurement?

    A.   Do not wait. Obtain the test immediately because the new levels of oxygen should be reflected immediately in arterial blood.

    B.   Wait 1 minute because that is about the amount of time it takes for one complete circulation of the blood.

    C.   Wait 3 minutes to account for the patient's lung abnormalities.

    D.   Wait 15 minutes to account for the patient's lung abnormalities.

2.   A patient with emphysema and an anxiety disorder comes to the emergency department complaining of shortness of breath. He was just told that one of his children was involved in an automobile accident and is clearly upset. He has no history of a recent cough or fever and was feeling well until he got the bad news. The chest radiograph shows no change from his prior radiograph, and his oxygen level is acceptable. He is breathing very rapidly, almost as he would if he were exercising. The cause of this patient's breathing discomfort is likely to be:

    A.   anxiety alone

    B.   emphysema alone

    C.   anxiety leading to physiologic changes in his breathing

    D.   physiologic changes from the emphysema leading to anxiety

3.   If a patient with emphysema breathes rapidly and does not allow sufficient time for exhalation, hyperinflation results. What physiologic principles prevent the lung from getting so big that it ruptures and causes a pneumothorax?

    A.   the stiffness of the chest wall

    B.   the compliance of the pleura

    C.   the tethering effect of the airways

    D.   the increased elastic recoil of the lung at higher lung volumes

    E.   the increased diameter of the airways at higher lung volumes

    F.   A and B

    G.   D and E

4.   You are caring for a patient who has pulmonary fibrosis, a condition that causes scarring in the lungs and is associated with reduced lung compliance. The TLC is 50% of predicted for the patient's age and height. What do you expect the peak or maximal expiratory flow to be?

   A.   normal
   B.   higher than normal
   C.   lower than normal
   D.   zero

5.   A patient with emphysema comes to the emergency department with acute respiratory distress. He is breathing at a rate of 28 breaths per minute and clearly appears to be laboring. He tells you that it is harder for him to breathe in than out despite the fact that his lung tests show significant expiratory airflow obstruction. Why does the patient complain more about inhalation than exhalation?

   A.   He has inspiratory obstruction as well.
   B.   He is hyperinflated.
   C.   He has pneumonia.
   D.   He has a neurological problem.
   E.   He does not understand the physiology of his disease.

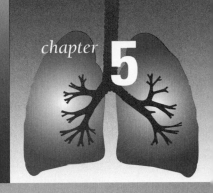

chapter 5

The Gas Exchanger: Matching Ventilation and Perfusion

CHAPTER OUTLINE

LEARNING OBJECTIVES

- To define the concepts of dead space and shunt.
- To describe the implications of changes in breathing pattern (e.g., rapid–shallow versus slow–deep breathing) on gas exchange.
- To outline the implications for gas exchange of the differences between the ways that oxygen and carbon dioxide bind with hemoglobin (specifically, the implications of a sigmoid-shaped versus a linear relationship between partial pressure of oxygen or carbon dioxide and hemoglobin binding).
- To describe the physiological implications for gas exchange of the anatomy of the pulmonary circulation (e.g., the relationship between pulmonary pressure and perfusion of alveoli, and the effect of hypoxemia on pulmonary blood flow).
- To define the physiologic causes of hypoxemia.
- To describe the principles that are used to distinguish the physiological causes of hypoxemia.
- To define the physiologic causes of hypercapnia.

Having examined in detail the physiological function of the ventilatory pump, we now turn our attention to what some might describe as the "business end" of the respiratory system, the gas exchanger. How does the system actually get oxygen into the blood and carbon dioxide into the alveolus? What are the effects of breathing pattern and body position on oxygen and carbon dioxide levels? Why do some physiologic derangements lead to low oxygen levels alone while others produce both low oxygen and high carbon dioxide levels in the blood? These are just some of the questions we examine in this chapter.

We start our study of the gas exchanger by looking at the distribution of air that is inhaled into the lungs and the impact of pleural pressure, body position, and breathing pattern on that distribution. Then we turn our attention to blood flow in the lungs and the factors that affect flow and resistance in the pulmonary circulation. To fully understand gas exchange, we must also investigate the different ways that carbon dioxide and oxygen bind to hemoglobin. These concepts have extremely important implications for the physiologic causes of low oxygen and high carbon dioxide levels in the blood.

Ultimately, gas exchange requires that we move air into and out of the alveoli (**ventilation**) and bring that air into contact with blood circulating through the lungs (**perfusion**). The matching of these two functions, or failure to do this efficiently, is one of the keys to our success or failure as aerobic beings.

Alveolar Ventilation and Carbon Dioxide Elimination

COMPONENTS OF VENTILATION: DEAD SPACE AND ALVEOLAR VENTILATION

Alveolar ventilation ($\dot{V}_A$) is a term used to characterize the volume of air per minute that enters or exits the alveoli of the lung. This gas is potentially able to participate in gas exchange, assuming that the alveoli are being supplied by blood (perfused). **Dead space ventilation** ($\dot{V}_D$), in contrast, is the term used to characterize the volume of air per minute that enters or exits the parts of the lung that do not participate in gas exchange. Dead space is composed of the volume of the conducting (non–gas-exchange) airways of the lung (**anatomic dead space**) as well as the volume of the alveoli that are not being perfused (**alveolar dead space**) (Fig. 5-1).

Total ventilation (usually represented as $\dot{V}_E$ with the E standing for *expired* gas because we usually measure ventilation by collecting the volume of expired gas over a fixed period of time) is the sum of alveolar ventilation and dead space ventilation:

$$\dot{V}_E = \dot{V}_A + \dot{V}_D$$

Anatomic dead space can be estimated in the average person as equal to 1 mL per pound of body weight. For the average person weighing 150 pounds, anatomic dead space would be 150 mL. Alveolar dead space in normal individuals is quite small, approximately 20 to 50 mL. This volume represents alveoli, typically in the apex of the lung in an upright person, that do not receive blood flow (more on the physiology of this phenomenon later in the chapter).

One of the ways to assess dead space and its likely impact on gas exchange is to express it as a proportion of the normal breath or **tidal volume** (V_T). A normal tidal volume in a person at rest is approximately 450 to 500 mL. If total dead space, sometimes called **physiological dead space**, is the sum of anatomic and alveolar dead space and these two elements add up to roughly 175 mL, then the **dead space to tidal volume ratio** (V_D/V_T) is 175/500, or approximately one third.

Measurement of Dead Space

Diseases of the lung often interfere with gas exchange. Our ability to characterize the severity of the disease relies, therefore, on the methods available to assess these derangements. The measurements of dead space and the ratio of dead space to tidal volume are used clinically for this purpose. For example, one definition of **respiratory failure**, the

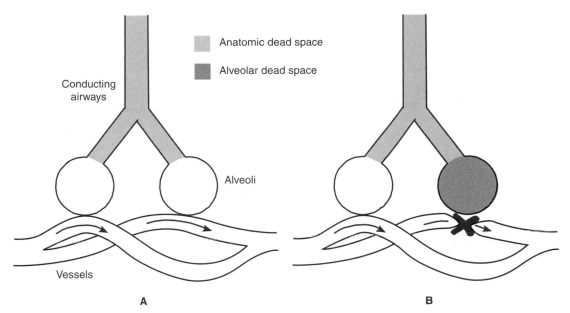

FIGURE 5-1 Dead space ventilation. **A,** A simple two-alveolar model of the lungs. Each of the alveoli is being perfused. Thus gas going to these alveoli participates in gas exchange. The air that enters the central airways and does not make it to the alveoli constitutes the anatomic dead space. This air does not participate in gas exchange. **B,** In this two-alveolar model of the lung, only one of the alveoli is being perfused. The other alveolus receives no blood flow. Thus, gas entering this alveolus is not in contact with blood and cannot participate in gas exchange. The volume of the alveoli that are ventilated but not perfused is termed the *alveolar dead space.*

inability of the respiratory system to sustain the metabolic needs of the body, is a V_D/V_T ratio that is greater than or equal to 0.6.

Anatomic Dead Space. As noted previously, anatomic dead space can be estimated from the size of the individual and is usually assumed to be equal to 1 mL/lb of body weight. Initial work on actually measuring anatomic dead space used the physiologic principles of gas exchange to develop appropriate methodologies.

Because the gas in the anatomic dead space at the end of an inhalation does not participate in gas exchange, it exits the mouth during exhalation with the same composition as it entered. The gas exiting the mouth after the anatomic dead space has been emptied represents gas from the alveoli. In normal individuals, most alveoli are perfused, but a small number are not; these nonperfused alveoli represent alveolar dead space. The gas breathed from the atmosphere contains approximately 79% nitrogen and 21% oxygen, with tiny amounts of carbon dioxide and other molecules. The gas exiting a perfused alveolus contains nitrogen (which is physiologically inert in the lung), oxygen from the atmosphere (not all of which is taken into the blood), and carbon dioxide. The carbon dioxide arrives at the alveolus carried by the blood returning from the body. It is displaced from hemoglobin by oxygen coming into the blood from the alveolus and then travels down a diffusion gradient into the alveolus.

Imagine now that you want to mark the gas coming from the anatomic or alveolar space, based on your knowledge of gas exchange, in order to measure the anatomic dead space. You could assess the concentration of carbon dioxide during an exhalation and plot the partial pressure of carbon dioxide as a function of the volume of exhaled gas (Fig. 5-2). Three phases can be observed. The initial gas exiting the mouth contains almost no

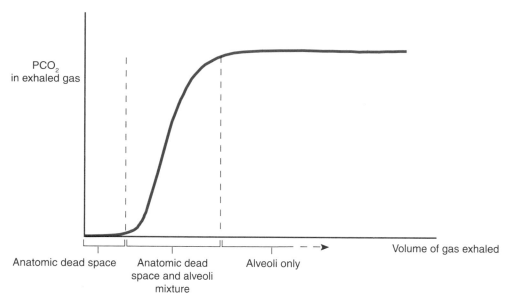

FIGURE 5-2 Measurement of anatomic dead space. The graph depicts the partial pressure of carbon dioxide in exhaled gas after a normal inhalation. The initial gas exhaled contains essentially no carbon dioxide. This gas represents the air in the anatomic dead space. The volume of gas associated with the increase in PCO_2 arises from a mixture of anatomic dead space and alveoli. The gas associated with the plateau in the PCO_2 arises from the alveoli.

carbon dioxide; this gas represents air coming from the anatomic dead space (where there is no gas exchange and hence almost no carbon dioxide). One then begins to see an increase in the carbon dioxide level. The gas now exiting the mouth is derived from a combination of the anatomic dead space and alveoli. Finally, the exhaled partial pressure of CO_2 reaches a plateau. This gas is assumed to be exiting solely from alveoli. By measuring the amount of gas exhaled that has essentially a PCO_2 of 0 mm Hg, one can estimate the anatomic dead space. A slightly more accurate estimate is achieved by measuring the volume of exhaled gas at the midpoint of the "transition zone," the portion of the graph when the PCO_2 is increasing, because this is a mixture of anatomic dead space gas and alveolar gas, as mentioned.

THOUGHT QUESTION 5-1: Imagine that you do not have equipment that would allow you to measure PCO_2, but you can measure the partial pressure of oxygen and nitrogen. How would you assess the anatomic dead space after asking the patient to inhale one breath of 100% oxygen?

Physiological Dead Space. Physiological dead space, as defined previously, is the sum of anatomic and alveolar dead space. It can be measured using the Bohr method, which is based on the following principle. If you collect exhaled gas in a bag from a person over several minutes, you have within the bag a combination of gas coming from perfused alveoli, nonperfused alveoli (alveolar dead space), and anatomic dead space. This type of collection is called **mixed expired gas**. If we measure the fraction of CO_2 in the gas, we can make the following assumption:

$$F_ECO_2 \times V_T = F_ICO_2 \times V_DCO_2 + F_ACO_2 \times V_A$$

$$F_ECO_2 = \text{the fractional amount of } CO_2 \text{ in mixed expired gas}$$

F_ICO_2 = the fractional amount of CO_2 in inhaled gas

F_ACO_2 = the fractional amount of CO_2 in alveolar gas

V_DCO_2 = volume of gas coming from the dead space (includes nonperfused alveoli)

V_T = tidal volume

V_A = alveolar volume (includes perfused alveoli only)

To state it simply, the fractional amount of CO_2 expired (which is equal to the partial pressure of the carbon dioxide divided by the barometric pressure, or $F_ECO_2 = PCO_2/Ptot$) multiplied by the exhaled volume (V_T) equals the amount of carbon dioxide leaving the lung. This amount must be equal to what is inhaled from the atmosphere and is sitting in the parts of the lung that are dead space and unaffected by gas exchange, added to the amount of carbon dioxide coming from alveoli that are perfused and are collecting carbon dioxide from the blood.

Because the fraction of carbon dioxide in inhaled gas (and hence, in the dead space) is essentially 0, the equation can be rewritten as follows:

$$F_ECO_2 \times V_T = F_ACO_2 \times V_A$$

Using the relationship that alveolar volume is the same as tidal volume minus dead space volume, the equation can be further transformed:

$$F_ECO_2 \times V_T = F_ACO_2 (V_T - V_DCO_2)$$

If we now transform fractional CO_2 into partial pressures by using the $FCO_2 = PCO_2/Ptot$ equation, we have:

$$P_FCO_2 \times V_T = P_ACO_2 \times V_T - P_ACO_2 \times V_D$$

(We have simply abbreviated VD_{CO_2} as V_D in the above equation.) Rearranging:

$$\frac{V_D}{V_T} - \frac{P_ACO_2 - P_ECO_2}{P_ACO_2}$$

Thus, the ratio of dead space to tidal volume is equal to the partial pressure of carbon dioxide in alveolar gas minus the partial pressure of carbon dioxide in mixed expired gas, divided by the partial pressure of carbon dioxide in alveolar gas. We can measure the partial pressure of CO_2 in mixed expired gas by collecting the expired gas, as previously noted.

In a normal person, the partial pressure of CO_2 in alveolar gas can be approximated by the PCO_2 at the very end of exhalation, called the *end-tidal CO2* (recall Figure 5-2 in our discussion of the measurement of anatomic dead space). In patients with significant lung disease, however, one does not always get a nice plateau in the expired CO_2 concentration because of abnormalities in the ventilation and perfusion of alveoli, and the measurement of end-tidal CO_2 can be misleading as a marker for alveolar CO_2. However, the blood in the capillaries that perfuse the alveoli quickly comes into equilibrium with the alveolar gas as CO_2 diffuses from pulmonary arterial blood into the air sacs the blood surrounds. Therefore, one may replace the P_ACO_2 with arterial partial pressure of carbon dioxide (P_aCO_2) as a reasonable approximation of the alveolar carbon dioxide. The P_aCO_2 level in a healthy person at rest is generally between 38 and 42 mm Hg. It is a common convention to use an uppercase *A* as the subscript indicating alveolar quantities and a lowercase *a* as the subscript indicating arterial quantities, and we follow this notation here.

CARBON DIOXIDE ELIMINATION: THE LUNG AS AN EXCRETORY ORGAN

Carbon dioxide is a byproduct of metabolism and, in high concentrations in the body, it can produce a number of toxic effects, including dyspnea (shortness of breath), severe acidosis, and altered levels of consciousness. The body's primary means of eliminating carbon dioxide is through the lungs via the alveoli. The excretory function of the lungs with respect to carbon dioxide can be summarized by the following relationship (note: K is a constant):

$$\dot{V}_A = K \times \frac{\dot{V}_{CO_2}}{P_aCO_2}$$

Let us examine the implications of this relationship. Alveolar ventilation, the amount of air going into and out of perfused alveoli, is directly proportional to carbon dioxide production ($\dot{V}_{CO_2}$); the greater the amount of carbon dioxide produced by the body, the greater the alveolar ventilation must be to maintain a constant partial pressure of carbon dioxide in the blood (P_aCO_2). If CO_2 production were to increase, for example, as is seen with fever, and alveolar ventilation did not increase, the P_aCO_2 would increase. Viewed from a different angle, if there were a problem with the ventilatory pump, such as a weakened diaphragm from an acute polio infection, that caused alveolar ventilation to be reduced while carbon dioxide production remained constant, then the P_aCO_2 would increase.

THOUGHT QUESTION 5-2: In what way might one view chronic hypercapnia (i.e., a P_aCO_2 > 40 mm Hg, with around 40 considered normal), as advantageous for a patient with a problem with the ventilatory pump and reduced ventilatory reserve (e.g., an individual with severe emphysema)?

The relationship between $\dot{V}CO_2$, alveolar ventilation, and the partial pressure of carbon dioxide in the arterial blood, as embodied by the equation above, may be considered a *clearance equation* for the lung. This relationship is analogous to the clearance equation used to describe kidney function in renal physiology (Table 5-1).

THOUGHT QUESTION 5-3: What you expect to happen to the P_aCO_2 in a patient with severe emphysema and essentially no respiratory system reserve capacity who is switched to a diet rich in carbohydrates? (Hint: recall from your knowledge of biochemistry the relative effects on carbon dioxide production of a diet rich in carbohydrates relative to one focused on proteins or fats.)

DETERMINANTS OF DISTRIBUTION OF VENTILATION

In the upright posture, ventilation is directed preferentially to the bases of the lungs. This is the consequence largely of the varying pleural pressures from the bases to the apices of the lungs. In an upright person, because of the impact of gravity on the lung and the mechanical interactions between the lungs and chest wall, pleural pressure is less negative at the base than at the apex of the lungs. For example, in a typical individual, pleural

TABLE 5-1 Clearance: Analogy Between The Physiology of the Lungs and Kidneys*

FOR THE LUNGS	FOR THE KIDNEYS
$$\dot{V}_A = K \times \dfrac{\dot{V}_{CO_2}}{P_aCO_2}$$	$$C_{creat} = \dfrac{U_{creat}V_u}{P_{creat}}$$

*The alveolar ventilation is analogous to the creatinine clearance of the kidneys. If the renal function were suddenly cut in half, the plasma creatinine level would double, assuming that the person was in a steady-state condition, making a constant amount of creatinine each day. Similarly, if respiratory system function were suddenly cut in half ($\dot{V}_A/2$), the P_aCO_2 would double if $\dot{V}_{CO_2}$ remained constant. C_{creat} = creatinine clearance; P_{creat} = concentration of creatinine in plasma; U_{creat} = concentration of creatinine in urine; $U_{creat}V_u$ = amount of creatinine produced in 24 hours; V_u = volume of urine in 24 hours.

pressure at the base at functional residual capacity (FRC) is -3 cm H_2O, and -8 cm H_2O at the apex. Consequently, at FRC, the alveoli at the apex are more distended than at the base. To understand this, recall the compliance relationship from Chapter 3: $C = \Delta V/\Delta P$. At FRC, the greater transmural pressure at the apex leads to larger alveoli there, assuming the compliance of the alveoli are the same throughout the lung (in other words, assuming all have the same compliance curve).

At this greater alveolar volume at FRC, the alveolus at the apex may now be on a flatter or less compliant portion of the pressure–volume curve than the alveolus at the base (Fig. 5-3). Thus, to get more air into this apical alveolus on the next inhalation (i.e., to expand it further) requires a greater change in pressure; the alveolus at the apex is less compliant than the alveolus at the base at FRC. Therefore, for a given change in pleural pressure resulting from activity of the inspiratory muscles, air will go preferentially to the more compliant alveoli at the bases.

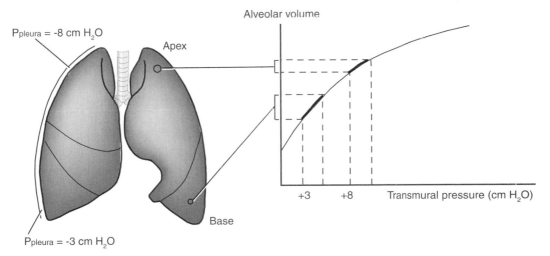

FIGURE 5-3 Pleural pressure and distribution of ventilation. At the lung volumes of normal resting breathing, apical alveoli exist on a flatter portion of the compliance curve compared with alveoli at the base. The alveoli near the apex may start at higher volumes when inspiration begins but undergo less of a change in volume for a given change in transmural pressure than the alveoli at the base. Changes in the volume of the apical and basal alveoli are shown on the vertical axis for identical changes in transmural pressure on the horizontal axis. It can be seen from this graph that ventilation is preferentially distributed to the bases; the change in volume of the basal alveoli is greater.

Physiology of the Pulmonary Vasculature: Where Does the Blood Go?

From Chapter 2, we recall that the pulmonary vasculature takes deoxygenated blood from the right ventricle to the alveolar capillaries, where gas exchange occurs. The capillaries rejoin to form pulmonary veins, which ultimately return the oxygenated blood to the left atrium and ventricle from where the oxygenated blood is pumped to the body. The pulmonary circulation is a high-compliance, low-resistance system. This means that it is able to adjust to large changes in flow, or cardiac output, with little change in resistance.

EFFECT OF GRAVITY ON PULMONARY BLOOD FLOW

Under conditions of normal cardiac output with an individual at rest, pulmonary blood flow is directed by gravity to the more dependent portions of the lung, that is, the portions of lung that are lower or oriented in the inferior aspects of the thorax with respect to gravity. Consequently, not all pulmonary vessels may receive blood and, thus, do not participate in gas exchange (recall that alveoli that do not receive blood flow constitute part of the dead space). Under conditions of high cardiac output or increased pulmonary vascular resistance, vessels that previously were not receiving blood are now recruited and participate in the gas exchange process. Nevertheless, as a general principle, the most dependent portions of the lung receive the greatest amount of pulmonary blood flow under normal conditions.

EFFECT OF THE ALVEOLI ON PULMONARY BLOOD FLOW

To the extent that pulmonary capillaries are in intimate contact with the alveoli, the pressure within the alveolus may alter the flow of blood through the adjacent capillary. If the alveolar pressure exceeds the pulmonary capillary pressure, the capillary will be squeezed and blood flow to that alveolus will diminish or cease.

Three types of pressure must be considered when assessing blood flow to a region of the lung: pulmonary arterial pressure, pulmonary venous pressure, and alveolar pressure (Quick Check 5-1). As discussed in Chapter 4, flow is normally expressed as a pressure differential divided by resistance:

$$\Delta P = Flow \times Resistance$$

$$\Delta P/Resistance = Flow$$

One might think that the pulmonary artery pressure (Pa) and the pulmonary venous pressure (Pv) are all that need to be considered when assessing flow through vessels with a given resistance. However, the pulmonary capillaries, as they course alongside the alveoli, appear to behave in a manner similar to the Starling resistors discussed in Chapter 4. As seen in Figure 4-10, if the vessel travels through an environment in which the pressure outside the vessel is greater than inside (i.e., a negative transmural pressure) and the vessel wall is very compliant, the vessel will collapse, and flow will cease (Fig. 5-4).

✓ QUICK CHECK 5-1   **FACTORS THAT AFFECT PULMONARY BLOOD FLOW**

Pulmonary arterial pressure
Pulmonary venous pressure
Alveolar pressure

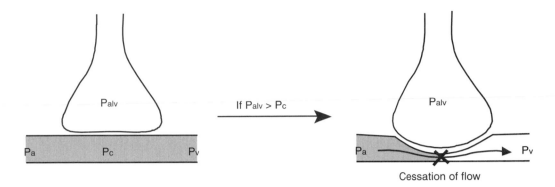

FIGURE 5-4 Pulmonary capillary as a Starling resistor. We depict a pulmonary artery leading to a capillary that interacts with an alveolus. Note that the pressure inside the blood vessel goes from P_a within the pulmonary artery to P_c within the pulmonary capillary and then to P_v within the pulmonary vein. If the pressure in the alveolus (P_{alv}) is greater than P_c, the capillary will narrow and collapse, thereby causing a cessation of blood flow to the alveolus. The alveolus then becomes part of the lung's dead space.

The relationship between these pressures and blood flow in the lungs can be characterized by thinking of the lungs as being composed of three zones, first postulated by John West (Fig. 5-5). In the most superior zone (highest region with respect to gravity), zone I, $P_{alv} > P_a$ and there is no flow of blood to this region of the lung. Thus, the alveoli in this location do not participate in gas exchange and are part of the lung's dead space. In zone II, $P_a > P_{alv} > P_v$. There is some blood flow to the alveolus throughout the zone, and the amount of flow increases as one moves from the top to the bottom of the lung and the pressure differential between Pa and Palv increases. In zone III, $P_a > P_v > P_{alv}$, and alveolar pressure has no effect on the flow of blood throughout the region.

PULMONARY VASCULAR RESISTANCE AND DISTRIBUTION OF FLOW

Because flow is inversely related to resistance, we also need to consider factors that will alter the resistance within the pulmonary circulation. Although the pulmonary arteries are relatively thin-walled structures, they do have smooth muscle within their walls. The tone of the muscle plays a role in modulating the diameter of the vessel and, hence, the resistance to flow. Unlike the systemic circulation, in which resistance is largely determined by the balance between the sympathetic and parasympathetic nervous systems, the pulmonary circulation does not appear to respond significantly to these inputs.

When exposed to low levels of oxygen, the pulmonary vessels constrict. The **exact** mechanism by which **hypoxic vasoconstriction** occurs is not well understood, but **hypoxia** is a very potent vasoconstrictor for the pulmonary circulation. Again, this is in contrast to the systemic circulation, which responds to localized tissue hypoxia with vasodilation. In the case of the systemic circulation, tissue hypoxia is often a sign of insufficient blood flow to that region; vasodilation in that setting makes sense as a way of delivering more blood and oxygen to sustain metabolic activities in the tissue. However, in the lungs, local hypoxia generally reflects a problem with the airways or alveoli: either ventilation is diminished because of obstruction of airflow or the alveoli are filled with fluid, thereby preventing gas from coming into contact with the pulmonary capillaries. In either event, it

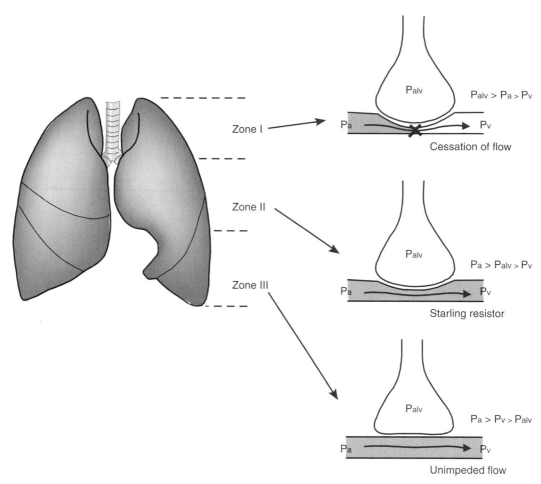

FIGURE 5-5 Zones of the lung. The lung is depicted as it would appear in a person standing upright. The most apical zone, zone I, of the lung receives the least blood flow because of gravity. $P_a < P_{alv}$ and the pulmonary vessel is compressed by the alveolus. In zone II, a gradual increase in flow occurs as one moves from the top to the bottom of the zone. P_a becomes increasingly large as one travels down the zone. Throughout the zone, however, $P_a > P_{alv} > P_v$. In zone three, the pressure in the alveolus has no impact on blood flow because $P_a > P_v > P_{alv}$.

would be counterproductive, from a survival standpoint, to send blood to regions of lung that are not receiving oxygen. The blood would exit that region of lung without having been able to eliminate carbon dioxide or pick up a fresh supply of oxygen. Rather, hypoxic vasoconstriction occurs, increasing local resistance in those vessels and resulting in a redistribution of pulmonary blood flow to the regions of lung that are well ventilated.

Two chemicals produced locally in the pulmonary circulation, nitric oxide (NO) and prostacyclin, also contribute to the tone and resistance of the pulmonary vessels. Nitric oxide, produced by nitric oxide synthase in the endothelial cell membrane, is the most potent endogenous vasodilator discovered. Its half-life is extremely short; it is produced locally and acts locally in the vessel. Mechanical stimuli can increase production of NO. Rapid flow through the vessel, for example, increases shear stress that appears to stimulate production of nitric oxide synthase. Thus, as flow increases, NO production also increases to dilate the pulmonary vessel and diminish resistance, thereby allowing for an increase in flow with a minimal increase in pressure. Biochemical stimulation of nitric oxide synthase results from a variety of materials, including acetylcholine, bradykinin, substance P, serotonin, and adenosine triphosphate (ATP).

Prostacyclins, a form of prostaglandin, are produced in the lungs and act as vasodilators in the pulmonary circulation. Also, to the extent that calcium influx into muscle cells is important for muscle contraction, pharmacologic blockage of calcium channels with drugs such as nifedipine and verapamil may lead to mild pulmonary vasodilatation.

MATCHING OF VENTILATION AND PERFUSION IN THE LUNGS

We have outlined the key physiological principles that govern the distribution of ventilation and perfusion in the lungs. For optimal gas exchange to occur, new gas from the atmosphere must travel to regions of lung that are also receiving blood flow from the pulmonary circulation. For the most part, both ventilation and perfusion are greatest at the bases of the lung in an upright person; thus, the match between ventilation and perfusion is generally quite good, although not perfect. There is relatively more ventilation than perfusion to the apices of the lungs, and relatively more perfusion than ventilation at the bases. When diseases affect either the distribution of ventilation ($\dot{V}$) or perfusion (Q), there is increased ventilation/perfusion mismatch ($\dot{V}$/Q mismatch), and gas exchange worsens. We return to this topic when we discuss causes of hypoxemia later in this chapter.

THOUGHT QUESTION 5-4: How does exercise affect ventilation/perfusion matching in the lung?

THOUGHT QUESTION 5-5: How would ventilation/perfusion matching change in outer space?

Carbon Dioxide Elimination Revisited: Joining Ventilation and Perfusion

Earlier in this chapter we described the relationship between alveolar ventilation and P_aCO_2. To appreciate fully the factors that affect this relationship, we must examine more closely the ways in which carbon dioxide is transported to and eliminated from the lungs.

CARBON DIOXIDE AND HEMOGLOBIN

Carbon dioxide is carried in the blood in three forms. First, CO_2 is bound to hemoglobin. Second, it is dissolved in the plasma component of blood. Finally, the dissolved CO_2 is in equilibrium with carbonic acid. Carbon dioxide and water combine to form carbonic acid, which dissociates to a proton and a molecule of bicarbonate.

$$CO_2 + H_2O \Leftrightarrow H_2CO_3 \Leftrightarrow H^+ + HCO_3^-$$

(In Chapters 6 and 7, we discuss this reaction and the importance of this relationship for the control of ventilation and the acid–base status of the body.) The dissolved carbon dioxide can be expressed as a partial pressure. The term **carbon dioxide content** in the blood, however, is reserved for the amount bound to hemoglobin plus the amount dissolved in the blood.

As blood travels from the lungs to the tissues, relatively little carbon dioxide is bound to hemoglobin. The PCO_2 of arterial blood in a normal individual ranges between 36 and 44 mm Hg. When the blood reaches the capillaries that perfuse metabolically active tissue, oxygen, which is in high concentration in the blood relative to the tissue, diffuses from the blood to support aerobic metabolism. Carbon dioxide, in high concentration in the tissues relative to the blood, diffuses from the tissues into the blood.

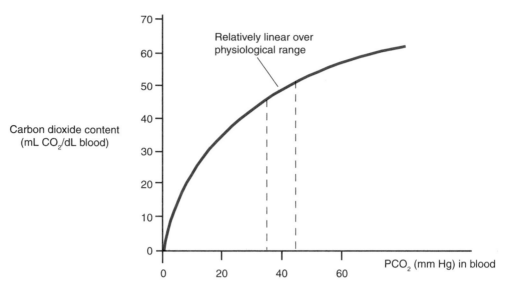

FIGURE 5-6 Carbon dioxide dissociation curve. Carbon dioxide content (the sum of the amount of carbon dioxide bound to hemoglobin and dissolved in plasma) is plotted as a function of the PCO_2 in the blood. Note that the relationship is relatively linear over the physiological range of PCO_2. **Also see Animated Figure 5-6.**

The relationship between carbon dioxide content in the blood and PCO_2 in the blood is relatively linear over the range found in the blood in humans (Fig 5-6). (It is common to see the term *content* used interchangeably with the more accurate term *concentration* when referring to the value for mL gas per 100 mL blood. You will see this for oxygen as well.)

The linear nature of this relationship makes it possible for the body to maintain a normal P_aCO_2, even in the presence of some poorly ventilated alveoli. If an alveolus is receiving little ventilation, the PCO_2 in the alveolus rapidly increases (carbon dioxide is diffusing from blood to the alveolus, but the gas in the alveolus is not being replaced with fresh gas from the atmosphere, which contains virtually no carbon dioxide). The result is that the diffusion gradient for carbon dioxide from the blood to the alveolus diminishes. Ultimately, the blood exiting the alveolus in the pulmonary vein has almost the same carbon dioxide content as the blood entering the alveolus. The P_aCO_2 increases and, as discussed in Chapter 6, stimulates the respiratory controller, which leads to an increase in ventilation as the body tries to compensate to restore P_aCO_2 to normal levels.

In the presence of some poorly ventilated alveoli, an increase in total ventilation restores P_aCO_2 to normal only if the carbon dioxide content of the blood leaving the normal alveoli can be lowered to a degree sufficient to compensate for the higher carbon dioxide content of the blood exiting the diseased alveoli. The linear relationship of the carbon dioxide curve makes this possible. Maximal ventilation of a normal alveolus can result in an alveolar PCO_2, and P_aCO_2, of 10 to 12 mm Hg. This process greatly reduces the carbon dioxide content of the blood perfusing this alveolus. When this blood mixes with blood from poorly ventilated alveoli, the final P_aCO_2 of the combined blood may be normal.

Use Animated Figure 5-6 to adjust the ventilation of the alveoli and observe what happens when you decrease ventilation for one alveolus and increase ventilation in another. Note that the well-ventilated alveolus can compensate for the poorly ventilated alveolus in terms of the CO_2 content of the outgoing mixed blood. As mentioned, this ability to compensate results from the linear relationship of the carbon dioxide dissociation curve, as seen in the diagram. Later in the chapter, we will see how the nonlinear shape of the oxygen dissociation curve creates a different situation for oxygen content when we mix blood from well-ventilated and poorly ventilated alveoli.

Hemoglobin contains four heme sites that bind oxygen and a protein chain onto which carbon dioxide binds. In the presence of high levels of carbon dioxide in the tissues, the increased binding to the protein chain alters the configuration of oxygen binding sites leading to release of oxygen to the tissue. When blood returns from metabolically active tissue to the right side of the heart, it has a high carbon dioxide content and low oxygen content. In the pulmonary capillary, the blood is exposed to a high PO_2 in the alveolus. The oxygen in the alveolus diffuses into the capillary blood, and carbon dioxide is released from hemoglobin; the hemoglobin binds oxygen in preference to carbon dioxide. The preferential binding of hemoglobin for oxygen and the resulting shift of carbon dioxide from being bound to hemoglobin to being dissolved in plasma is manifest as a shift to the right in the carbon dioxide-hemoglobin dissociation curve.

This shift in the curve is termed the *Haldane effect* (Fig. 5-7). Note that for any given level of carbon dioxide content, the P_aCO_2 is higher in the presence of a high concentration of oxygen. If, because of a problem with the respiratory controller or the ventilatory pump, the lungs are not able to increase alveolar ventilation to eliminate the increased amount of dissolved carbon dioxide, the P_aCO_2 will increase. This is one of the reasons, for example, that a patient with emphysema may have an increase in P_aCO_2 with administration of supplemental oxygen.

PHYSIOLOGICAL CAUSES OF HYPERCAPNIA

There are four basic physiological causes of hypercapnia (i.e., elevated P_aCO_2), and they can all be derived from a basic relationship discussed earlier in this chapter:

$$\dot{V}_A = K \times \frac{\dot{V}_{CO_2}}{P_aCO_2}$$

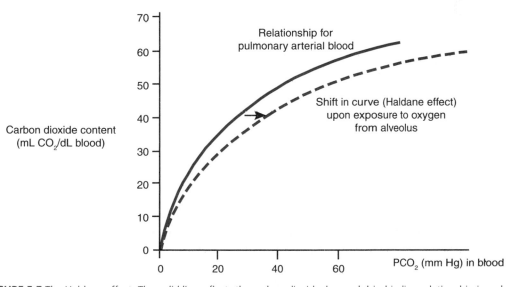

FIGURE 5-7 The Haldane effect. The *solid line* reflects the carbon dioxide–hemoglobin binding relationship in pulmonary artery blood. As the blood enters the pulmonary capillary and comes into contact with an alveolus with a high PO_2, oxygen diffuses from the alveolus into the blood and binds preferentially with the hemoglobin. Carbon dioxide is displaced from the hemoglobin and enters the plasma. The curve is shifted to the right (*dashed line*). For any given level of carbon dioxide content, the PCO_2 is now higher.

If alveolar ventilation ($\dot{V}_A$) is reduced and carbon dioxide production ($\dot{V}_{CO2}$) remains constant, the P_aCO_2 must increase. Three physiologic changes may result in a decrease in alveolar ventilation: (1) a decrease in total or minute ventilation ($\dot{V}_E$); (2) an increase in dead space ventilation attributable to a rapid, shallow breathing pattern; and (3) an increase in dead space caused by $\dot{V}/\dot{Q}$ mismatch in the context of a constant $\dot{V}_E$. The fourth physiological cause of hypercapnia is an increase in carbon dioxide production in the context of a patient who is unable to compensate by increasing ventilation, either because of a controller problem (e.g., a drug overdose that suppresses the respiratory controller) or an abnormality of the ventilatory pump, as is seen with severe emphysema.

| QUICK CHECK 5-2 | PHYSIOLOGICAL CAUSES OF HYPERCAPNIA |

Decreased minute ventilation
Decreased alveolar ventilation caused by a rapid, shallow breathing pattern
Ventilation/perfusion mismatch
Increased carbon dioxide production in the setting of a fixed ventilation

Ventilation/perfusion mismatch, the result of diseases of the gas exchanger that lead to a disproportionate amount of blood going to relatively poorly ventilated alveoli, is a common finding, but the mismatch causes hypercapnia relatively infrequently. As the P_aCO_2 begins to increase, the body's usual response is to increase ventilation (more on this in Chapter 6). As the well-ventilated alveoli receive an increase in the amount of gas entering and exiting each minute, the PCO_2 in these lung units may decrease well below normal levels. Based on the carbon dioxide–hemoglobin dissociation curve described previously, the blood perfusing those units leaves the alveoli with a very low carbon dioxide content and compensates for the high carbon dioxide level in the blood coming from the poorly

 ventilated alveoli (see Animated Figure 5-6).

Oxygenation and the Gas Exchanger

Having examined the factors that affect carbon dioxide elimination from the blood, we now turn our attention to oxygen. The binding of oxygen to hemoglobin is quite different than for carbon dioxide. These differences have a profound effect on the physiologic causes of hypoxemia and the response of the body to breathing supplemental oxygen in various disease states.

OXYGEN BINDING TO HEMOGLOBIN

In contrast to the carbon dioxide–hemoglobin dissociation curve, which is linear, the oxygen–hemoglobin dissociation curve is sigmoid or S shaped (Fig. 5-8). The y axis denotes the percentage of hemoglobin that is in the oxy-hemoglobin state, that is, the oxygen-binding sites on hemoglobin are filled with oxygen atoms. This is expressed as a function of the partial pressure of oxygen in the plasma.

The shape of the curve facilitates the release of oxygen in hypoxic tissues and the uptake of oxygen by the blood in the alveolus. The PO_2 of the tissues is typically about 40 mm Hg. At this partial pressure, oxygen leaves hemoglobin (note the steep slope of the curve as the P_aO_2 decreases below 60 mm Hg), as reflected by the lower saturation on the vertical axis.

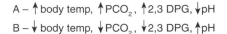

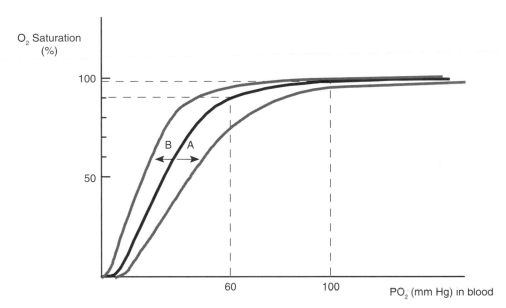

FIGURE 5-8 The oxygen–hemoglobin dissociation curve. The oxygen saturation (the percentage of hemoglobin in the oxy-hemoglobin state) is plotted as a function of PO_2. Note the sigmoid shape of the curve. Above a partial pressure of about 60 mm Hg, the curve is relatively flat, with oxygen saturations above 90%. Below a PO_2 of 60 mm Hg, the oxygen saturation decreases rapidly (i.e., the oxygen comes off the hemoglobin). The PO_2 of 100 mm Hg shown on the graph corresponds to a normal alveolar PO_2 (at sea level) and translates to an oxygen saturation of nearly 100%. Factors that shift the curve to the right include increased body temperature, increased PCO_2, increased 2,3 DPG, and decreased pH. Factors that shift the curve to the left include decreased body temperature, decreased PCO_2, decreased 2,3 DPG, and increased pH. **Also see Animated Figures 5-8A and 5-8D.**

Thus, as arterial blood travels through the capillaries in the tissues, the oxygen is released to support aerobic metabolism.

When the venous blood returns to the alveoli, it is exposed to a PO_2 of approximately 100 mm Hg in the alveolus (more on this in a moment). Oxygen diffuses from the alveoli into the plasma, displaces carbon dioxide from the hemoglobin (remember the Haldane effect), and binds to the hemoglobin until the saturation reaches the value that corresponds to the PO_2 in the plasma.

The relatively flat portion of the curve between a P_aO_2 of 60 and 100 mm Hg has another evolutionary advantage: it ensures that the hemoglobin will remain nearly fully saturated with oxygen even when the alveolar—and hence, arterial, PO_2—decreases to levels as low as 60 mm Hg. Because most of the oxygen carried in the blood is bound to hemoglobin, this ensures that the **oxygen content** of the blood (the sum of the oxygen bound to hemoglobin and dissolved in the blood) will remain high. Thus, tissues receive adequate amounts of oxygen when a person is at moderate altitudes or when there is a problem with the gas exchanger, such as pneumonia, that causes the P_aO_2 to decrease below 100 mm Hg.

When we examined the carbon dioxide–hemoglobin dissociation curve, we saw that the relationship could be altered by the addition of oxygen; the curve shifted to the right. The oxygen–hemoglobin relationship may also be altered by a number of factors that increase the affinity of hemoglobin for oxygen (shift in the curve to the left) or decrease the affinity for oxygen (shift in the curve to the right). When the pH of the blood decreases (i.e., the hydrogen ion concentration increases), the curve is shifted to the right, a phenomenon termed the Bohr effect (and conversely, an increase in pH shifts the curve to the left). Hydrogen ion binds to hemoglobin and alters the conformation of the oxygen–heme binding sites, much as carbon dioxide does, to reduce the affinity of hemoglobin for oxygen. Because a decrease in pH in the tissues often reflects anaerobic metabolism, the product of which is lactic acid, the shift of the curve to the right may be seen as a compensation that results in greater release of oxygen to the tissues (for a given PO_2, less oxygen is bound to hemoglobin and more is available to the tissue) and an attempt by the system to restore aerobic metabolism. Temperature affects the relationship (increases in body temperature shift the curve to the right), as do carbon dioxide (increases in $PaCO_2$ shift the curve to the right) and 2,3 diphosphoglycerate, an organic phosphate that is a byproduct of red blood cell (RBC) metabolism (a decrease in 2,3 DPG shifts the curve to the left).

 Use Animated Figure 5-8A to observe the effects of shifts in the oxygen dissociation curve. The horizontal axis shows PO_2, including values corresponding to the alveoli and the tissues. The vertical axes show oxygen saturation of hemoglobin and the oxygen concentration of the blood. The shaded portions of the vertical axes show the differences in oxygen saturation and content between the alveolus and the tissues (in other words, how much oxygen is offloaded). Note how shifting the curve to the right facilitates oxygen offloading at the tissues (shown by the increase in the saturation and content differences for oxygen between alveolus and tissues). In particular, as you shift the curve to the right, notice how the alveolar PO_2 still corresponds to the flat portion of the dissociation curve (and thus a high oxygen saturation), but the saturation corresponding to the tissues' PO_2 (located on the steep portion of the curve) decreases quite a bit as the curve shifts.

THOUGHT QUESTION 5-6: What do the factors that affect the oxygen–hemoglobin dissociation curve have in common when one considers the things that cause the curve to shift to the right?

Carbon monoxide (CO), a byproduct of combustion, is a toxic gas that exerts its effects by binding to hemoglobin. When present in the environment, CO competes with oxygen for the heme-binding sites on hemoglobin (hemoglobin's affinity for CO is 200 times greater than for oxygen), thereby reducing the amount of oxygen carried by the blood to the tissues. In addition, CO alters the relationship between PO_2 and oxygen saturation. The oxygen–hemoglobin saturation curve loses its sigmoid shape and is shifted to the left (Fig. 5-9).

This change in the curve reflects the greater affinity of oxygen to hemoglobin in the presence of CO and greatly accentuates the clinical consequences of CO poisoning. The consequence is that less oxygen is released to the tissues than would be predicted based solely on the reduced oxygen content of the blood. The effect of the combination of these two factors is to deprive metabolically active tissues of oxygen. Carbon monoxide poisoning is characterized by alterations in cognitive function, seizures, coma (in severe cases), chest pain, and the accumulation of metabolic acid in the blood caused by the body's necessity to use anaerobic metabolism to generate energy.

OXYGENATION

In this section we investigate the physiological causes of low oxygen levels in the blood. Hypoxia is a generic term that refers to a low PO_2. **Hypoxemia** refers to a low PO_2 in the blood.

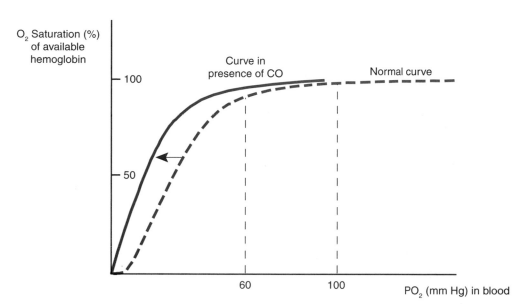

FIGURE 5-9 Carbon monoxide (CO) and hemoglobin. When CO binds to hemoglobin, the oxygen–hemoglobin dissociation curve loses its sigmoid shape and shifts to the left. This is the result of a conformational change in the hemoglobin that is the consequence of the binding with CO. As a result of the shift in the curve, oxygen is bound more tightly to the hemoglobin, less oxygen is released to the tissues, and aerobic metabolism is impaired; this process is functionally equivalent to depriving the tissues of oxygen.

The Alveolar Gas Equation

The **alveolar gas equation** is used to calculate the partial pressure of oxygen within the ideal alveolus in the lung (P_AO_2; recall that a lower case a signifies arterial gas and a capital A denotes alveolar gas). ("Ideal" is used here to mean that this would be the value seen in all alveoli if there were no $\dot{V}/Q$ mismatch.) The equation describes the relationship between the concentration of inspired oxygen, the ventilation of the alveolus (as represented by the alveolar carbon dioxide, or P_ACO_2), and the P_AO_2.

$$P_AO_2 = FIO_2 (P_{atm} - P_{H_2O}) - P_ACO_2/R$$

- FIO_2 = the fraction of oxygen in the inspired gas (for atmospheric gas, this number is 0.21)
- Patm = barometric pressure (remember, this changes depending on altitude), which is 760 mm Hg at sea level
- P_{H_2O} = water vapor pressure when the gas is fully saturated (because the air we breathe is humidified by the upper airway on its way to the alveolus), which is equal to 47 mm Hg
- R = the **respiratory quotient** which represents the ratio between oxygen consumed and carbon dioxide produced by the body (as oxygen is absorbed from the alveolus, carbon dioxide is eliminated proportional to this ratio). The respiratory quotient varies with the nature of the individual's diet. In a typical American diet, R = 0.8. If an individual eats primarily carbohydrates, R becomes closer to 1.0. If an individual eats a diet rich in fat, R becomes closer to 0.7.

At sea level, for an individual with a normal P_aCO_2 of 40 mm Hg (because carbon dioxide diffuses readily between the blood and the alveolus, one can assume that alveolar and arterial carbon dioxide have come into equilibrium by the time the blood leaves the pulmonary capillary), the P_AO_2 is approximately 100 mm Hg ($0.21(760 - 47) - 40/0.8 = 100$).

To determine the P_AO_2, you must obtain a sample of arterial blood, measure the P_aO_2 and P_aCO_2 (the sample is called an **arterial blood gas [ABG]**, which provides you with the P_aO_2, P_aCO_2, and pH of the blood), and compute the P_AO_2 from the alveolar gas equation. With this value in hand, you can now compare the alveolar and arterial PO_2. Because ventilation and perfusion are not perfectly matched, even in normal individuals (remember that there is relatively more ventilation toward the apex of the lungs and relatively more perfusion toward the bases), there is always a difference between P_AO_2 and P_aO_2. In addition, the normal gradient may reflect the mixture of a small amount of deoxygenated blood from the bronchial veins, with oxygenated blood returning from the lungs to the left heart. The difference between P_AO_2 and P_aO_2 is called the **alveolar to arterial oxygen difference (A-aDO$_2$)** and defines whether there is a problem with the gas exchanger. The A-aDO$_2$ increases in normal people as they age because of changes in ventilation that relate to loss of elastic recoil in the lung. Until age 30 years, a normal A-aDO$_2$ is less than or equal to 10 mm Hg. At ages older than 30 years, the normal A-aDO$_2$ can be approximated as age $\times$ 0.3 (e.g., a 70-year-old person can have an A-aDO$_2$ of 20 mm Hg and still have a normal gas exchanger). With supplemental oxygen, the A-aDO$_2$ increases because of mild worsening of $\dot{V}/Q$ mismatch and an increased effect on P_aO_2 of the small amount of shunted blood (deoxygenated blood returning to the *left* side of the heart) under these conditions (a full discussion on the effect of shunt on P_aO_2 will follow shortly). Because it is difficult to know what constitutes a normal A-aDO$_2$ when a patient is breathing supplemental oxygen, we primarily perform this calculation based on the patient breathing room air.

If a person has a low P_aO_2, it is important to calculate the A-aDO$_2$ to determine if the hypoxemia is caused by a problem with the gas exchanger. If the A-aDO$_2$ is abnormally high, then there must be a pathological problem with either the lung tissue or the pulmonary circulation. If the A-aDO$_2$ is normal, the gas exchanger (the alveoli and pulmonary capillaries) is normal and an alternative explanation must be sought to explain the hypoxemia.

Causes of Hypoxemia

Alveolar Hypoventilation. If alveolar ventilation is reduced due to a reduction in total ventilation or a change in the pattern of breathing (e.g., more rapid, shallow breaths), the P_ACO_2 increases because the gas in the alveolus is not being exchanged at the normal rate. Thus, carbon dioxide delivered from the blood perfusing the alveolus will begin to accumulate. As you look back at the alveolar gas equation, you will see that an increase in the P_ACO_2 causes the P_AO_2 to decrease. Physiologically, the decrease in alveolar ventilation leads to a decrease in alveolar oxygen because the oxygen present in the alveolus is diffusing into the blood but is not being replaced at a normal rate. The alveolar and arterial partial pressure of oxygen will decrease by the same amount. The A-aDO$_2$ will remain within the normal range. Thus, one can conclude that the hypoxemia is not caused by a gas exchanger problem. In this case, the hypoxemia would be caused by an abnormality of the controller (see Chapter 6) or a neuromuscular problem, such as myasthenia gravis (a disease that interferes with the transmission of neural impulses to skeletal muscles), that affects the ventilatory pump.

Reduced PIO$_2$. When you go to a high altitude, the P_aO_2 decreases. Looking at the alveolar gas equation, the P_AO_2 is affected at higher altitudes because barometric pressure, Patm, is reduced. The FIO$_2$ remains the same, but with a reduction in barometric pressure, the partial pressure of the inspired gas decreases. As with decreases in alveolar ventilation, the decrease of the alveolar and arterial partial pressure of oxygen will be the same, and the A-aDO$_2$ will remain within the normal range. The hypoxemia associated with altitude, therefore, is not caused by a problem with the gas exchanger. A similar decrease in the PIO$_2$ might be seen when the FIO$_2$ is reduced. A person who was caught in an enclosed space in which a fire is burning will be exposed to a low FIO$_2$ because of the consumption of oxygen by the fire.

Ventilation–Perfusion Mismatch. As noted previously, even normal individuals have a small amount of V̇/Q mismatch that leads to a discrepancy between alveolar and arterial oxygenation. Many disease states, including pneumonia, heart failure, asthma, and emphysema, can cause a worsening of V̇/Q mismatch. The hypoxemia seen in these conditions, if caused by V̇/Q mismatch, is associated with an abnormally large A-aDO$_2$. V̇/Q mismatch is the most common pathologic explanation for hypoxemia.

Earlier in the chapter, we observed how well-ventilated alveoli can compensate for poorly ventilated alveoli in terms of the CO$_2$ content of the outgoing mixed blood (see Animated Figure 5-6). At this point, we examine how the nonlinear shape of the oxygen dissociation curve creates a different situation for oxygen content when we mix blood from well-ventilated and poorly ventilated alveoli, as occurs in V̇/Q mismatch.

 First, use Animated Figure 5-8B to vary the V̇/Q ratio for a single alveolus, and observe how the PO$_2$ and oxygen content of the outgoing capillary blood change. In particular, note that a low V̇/Q ratio produces a PO$_2$ value that corresponds to the steep portion of the oxygen dissociation curve and thus significantly lowers the oxygen content of the outgoing capillary blood. In contrast, a high V̇/Q ratio produces a PO$_2$ value that corresponds to the relatively flat portion of the oxygen dissociation curve and thus does not significantly elevate the oxygen content of the outgoing blood compared with a normal V̇/Q ratio.

 Keeping this in mind, now use Animated Figure 5-8C to vary the V̇/Q ratios for a three alveolus model, and observe the changes in the oxygen content of the outgoing mixed blood. Try increasing the V̇/Q ratio in one alveolus while decreasing the ratio in another, and note how the small elevation in oxygen content caused by the high V̇/Q ratio alveolus does not compensate for the significant decrease in oxygen content caused by the low V̇/Q ratio alveolus. This observation is one of the keys to understanding how worsening V̇/Q mismatch may lead to hypoxemia.

An extreme form of V̇/Q mismatch occurs when some of the alveoli of the lung completely collapse or are filled with fluid. In these cases, alveolar ventilation in the affected regions of the lung is 0. We term this condition **shunt**. (Note that even though shunt is an extreme form of V̇/Q mismatch, we distinguish these terms when referring to causes of hypoxemia and typically use V̇/Q mismatch to refer to the less extreme non-shunt case.) Although you would normally expect the blood vessel serving a collapsed or fluid-filled alveolus to constrict because of local hypoxia, the disease process causing the alveolar process may also affect the pulmonary circulation such that perfusion of the nonventilated alveolus continues. Thus, the shunt is established. Hypoxemia caused by a shunt, similar to hypoxemia caused by other causes of V̇/Q mismatch, is characterized by an abnormally large A-aDO$_2$, indicating a problem with the gas exchanger.

Diffusion Abnormality. The movement of gas between the alveolus and the blood occurs by diffusion. Both oxygen and carbon dioxide diffuse quickly so that under normal conditions in a patient at rest, equilibration is reached well before an RBC completes its travel through the alveolar capillary (carbon dioxide diffuses approximately 20 times more rapidly than does oxygen and is able to establish an equilibrium between the blood and the alveolus despite a relatively low diffusion pressure). For oxygen, equilibration occurs when the RBC is approximately one third of the way through the capillary.

As a result of this rapid equilibration, significant "reserve capacity" is available for diffusion. In other words, even if a disease process interferes with diffusion and equilibration takes twice as long, the blood will have picked up all of its oxygen before the RBC exits the capillary. Thus, diffusion abnormalities are not usually a cause of hypoxemia at rest.

During exercise, cardiac output may increase by fivefold. This increase in blood flow leads to more rapid transit of the RBCs through the alveolar capillary. With exercise, if there is a problem with diffusion, the reserve that was present at rest no longer exists, and hypoxemia may develop.

In Animated Figure 5-10, equilibration of oxygen is shown along the pulmonary capillary. Observe how equilibration takes place farther along the capillary with increasing flow, as occurs during exercise. As mentioned, equilibration may not occur if conditions exist in which flow is sufficiently increased and a concurrent problem with diffusion is present.

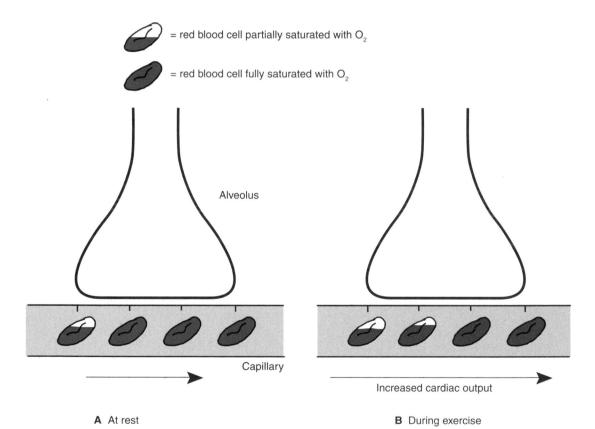

FIGURE 5-10 Diffusion of oxygen. **A,** At rest, equilibration between the alveolus and the blood occurs quickly. By the time a red blood cell (RBC) is one third of the way through the alveolar capillary, the hemoglobin has picked up all the oxygen with which it is going to bind. **B,** During exercise, however, cardiac output increases, resulting in more rapid transit of RBCs through the capillary (Fig. 5-10). In the diagram, equilibration between alveolus and blood is now shown as occurring farther along the capillary. If a pathologic process is present in the lung that prolongs diffusion time, equilibration may not occur between the alveolus and the blood. The result is a decrease in P_aO_2 during exercise. Also see Animated Figure 5-10.

Oxygen Content

Oxygen exists in the blood in two forms: bound to hemoglobin and dissolved in the liquid portion of the blood. The oxygen content of the blood reflects the amount of oxygen that is bound to hemoglobin and the amount dissolved in the liquid portion, or plasma component, of the blood (i.e., the total amount of oxygen in the blood). The equation for calculation of oxygen content is:

$$O_2 \text{ content} = \text{Amount } O_2 \text{ bound to Hgb} + \text{Amount } O_2 \text{ dissolved in blood}$$
$$O_2 \text{ content} = 1.35(\text{Hgb})(\text{Oxygen saturation}) + 0.003 \ (P_aO_2)$$

Each gram of hemoglobin can combine with approximately 1.35 mL of oxygen (you may see values of 1.34 to 1.39 mL of oxygen per gram of hemoglobin in different references). A normal hemoglobin value is approximately 14 g/100 mL of blood. The constant 0.003 transforms the P_aO_2 into mL of oxygen per 100 mL of blood.

Assuming a P_aO_2 of 100 mm Hg and a corresponding oxygen saturation of approximately 97.5% as well as a normal hemoglobin level, the amount of oxygen bound to hemoglobin is 18.4 mL/100 mL of blood. The amount of oxygen dissolved in the blood is 0.3 mL/100 mL of blood. As you can see, the amount of oxygen bound to hemoglobin overwhelms the amount of oxygen dissolved in the blood. Furthermore, as soon as the hemoglobin is fully saturated or nearly fully saturated, further increases in P_aO_2 make relatively little difference in the oxygen content of the blood (recall Animated Figure 5-8B). For example, in the illustration cited above, if we give the person supplemental oxygen and increase the FIO$_2$ (fraction of inspired oxygen) from 0.21 to 0.35, the P_AO_2 will be 200 mm Hg. Assuming an A-aDO$_2$ of 25 mm Hg (note that this value is increased from when the patient is on room air), the P_aO_2 will be 175 mm Hg. With this increase in the P_aO_2 (and a new corresponding oxygen saturation of 99%), the new oxygen content is only 18.7 + 0.5 = 19.2 mL/100 mL of blood, barely above the 18.4 + 0.3 = 18.7 mL/100 mL of blood when the person was breathing room air.

A low hemoglobin level, or **anemia**, reduces oxygen content significantly. It does not change the P_aO_2, however. Thus, a person with anemia may have symptoms secondary to reduced delivery of oxygen to metabolically active tissues even though no hypoxemia (which refers specifically to a low PO$_2$ in the blood) is present. Use Animated Figure 5-8D to vary the hemoglobin level and observe the effect on the oxygen–hemoglobin dissociation curve. Notice that the saturation level stays the same for any given PO$_2$ (as shown by the saturation axis that scales with the amount of hemoglobin), but the total content (or concentration) of oxygen delivered varies with changes in the amount of hemoglobin (as shown by the changing O$_2$ concentration difference on the vertical axis).

Response to Supplemental Oxygen: Distinguishing Shunt from Less Extreme Ventilation–Perfusion Mismatch

In a patient who has hypoxemia due to $\dot{V}$/Q mismatch, some areas of the lung have relatively poor ventilation compared with the amount of blood perfusing the alveolus. Oxygen diffuses from the alveolus to the blood but is not quickly replaced because ventilation of the lung unit is reduced. The P_AO_2 decreases. Subsequent blood that enters the alveolar capillary is exposed to a low P_AO_2 and exits the capillary with a low P_aO_2. If one now places the patient on supplemental oxygen and increases the FIO$_2$, the alveolar oxygen will increase (even though ventilation of the alveolus has not changed, P_AO_2 will increase because the gradient for diffusion of oxygen between the airway and the alveolus has increased). Although there is an increase in A-aDO$_2$ when on supplemental oxygen (as mentioned in our discussion of the alveolar gas equation), it is relatively small, and the P_aO_2 increases almost as much as the P_AO_2. Thus, with administration of supplemental oxygen to a patient who is hypoxemic because of $\dot{V}/Q$ mismatch, the P_aO_2 will increase, as will oxygen content.

Now consider the extreme of $\dot{V}/Q$ mismatch that we call *shunt*, when the ventilation of an alveolus is 0. The blood perfusing the alveolus sees no gas and exits the alveolar capillary with the same PO_2 as it entered the alveolus. If one administers supplemental oxygen, although FIO_2 is increased, the blood perfusing the nonventilated alveolus will still not see any gas and will exit the alveolar capillary without picking up any oxygen. What about the alveoli that are being ventilated? The P_AO_2 of these alveoli increases as FIO_2 is increased, but the oxygen content of the blood perfusing these alveoli does not change significantly. The hemoglobin exiting these normal alveoli was nearly fully saturated when the patient was breathing room air. Based on what you have learned about the calculation of oxygen content, the increase of P_AO_2 under these circumstances will have relatively little effect on the oxygen content (Fig. 5-11).

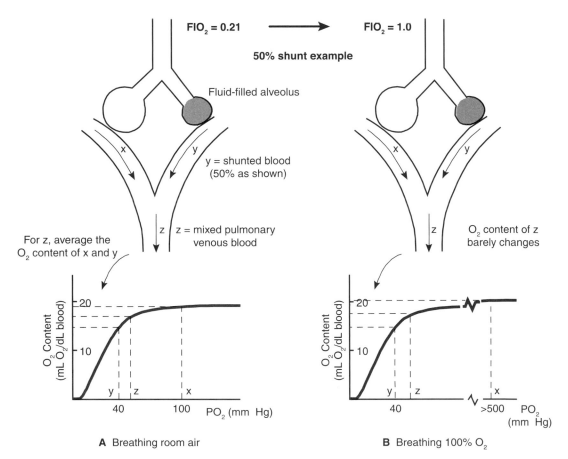

FIGURE 5-11 Supplemental oxygen and shunt. **A,** A two-alveolar model of the lung is depicted with one alveolus receiving normal ventilation and one half of the body's perfusion. The other alveolus receives no ventilation and one half of the perfusion (i.e., it represents a shunt). With the patient breathing room air, a significant degree of hypoxemia is present. **B,** Now the patient has been given supplemental oxygen with $FIO_2 = 1.0$, and the oxygen content is recalculated. Despite the tremendous increase in FIO_2, the oxygen content of the blood exiting the lung (normal alveolus plus shunt) has increased by less than 5%. **Also see Animated Figure 5-11.**

When one mixes the blood from the well-ventilated alveoli with blood from the areas of shunt, the final oxygen content of the blood is changed minimally after the addition of supplemental oxygen. Animated Figure 5-11 illustrates this effect. Vary the degree of shunt and compare the final oxygen content of the mixed outgoing blood for the patient breathing room air versus 100% oxygen. Note in particular where the PO_2 values of the shunted, nonshunted, and mixed outgoing blood fall on the oxygen–hemoglobin dissociation curve. The PO_2 of the nonshunted blood corresponds to the relatively flat portion of the dissociation curve, whether breathing room air or 100% O_2. Thus, the incremental gains in O_2 content in the nonshunted blood from breathing 100% O_2 make little difference to the O_2 content of the mixed outgoing blood (as shown on the vertical axis of graph).

The response of the P_aO_2 to supplemental oxygen allows one to distinguish clinically, at the bedside, hypoxemia caused by $\dot{V}/Q$ mismatch (significant increase in P_aO_2) from hypoxemia caused by shunt (no significant increase in P_aO_2).

THOUGHT QUESTION 5-7: If a patient has hypoxemia caused by a right-to-left shunt (in other words, blood coming directly from the right side to the left side of the heart without passing through the lungs) in the heart because of an atrial septal defect (i.e., hole in the wall separating the right and left atria), what would you expect the response of the P_aO_2 to be when the patient is given supplemental oxygen?

Distinguishing the Physiological Causes of Hypoxemia

We have now outlined the five physiological causes of hypoxemia: reduced PIO_2, hypoventilation (decreased alveolar ventilation), $\dot{V}/Q$ mismatch, shunt, and diffusion abnormalities (note: we are not talking about specific disease states but rather the physiology that underlies conditions, both normal [e.g., high altitude] and pathological [e.g., pneumonia, emphysema]). Although shunt is actually an extreme form of $\dot{V}/Q$ mismatch, we list it separately for two reasons. First, the diseases that lead to significant shunt are relatively few in number and very distinct, for example, the adult respiratory distress syndrome and septal defects in the heart. Second, shunt can be distinguished very easily in the clinical setting from all other conditions that lead to less severe forms of $\dot{V}/Q$ mismatch. In a patient you suspect has $\dot{V}/Q$ mismatch or shunt, you should check the initial P_aO_2, administer supplemental oxygen, and recheck the P_aO_2. If a significant increase in the P_aO_2 has occurred with supplemental oxygen, one is dealing with typical $\dot{V}/Q$ mismatch; if the P_aO_2 does not change significantly, the patient has a shunt that accounts for the hypoxia (the appropriate magnitude of change in P_aO_2 for administration of 100% O_2 with different degrees of shunt can be appreciated by using Animated Figure 5-11).

In considering the physiology of a patient with hypoxemia, you should adopt a systematic approach to your analysis. First, calculate the A-aDO$_2$. This must be done with the patient breathing room air (i.e., off supplemental oxygen). With most techniques we use to administer supplemental oxygen, with the exception of the patient in respiratory failure who has an endotracheal tube in place and is attached to a ventilator, it is difficult to know the exact FIO_2. Without a well-defined FIO_2, it is impossible to calculate accurately the A-aDO$_2$ in a patient receiving oxygen via a mask or nasal tubing. Calculation of the A-aDO$_2$ in some patients receiving supplemental oxygen is complicated by one other factor. In patients with obstructive lung disease who have been receiving supplemental oxygen, it is important to remove the oxygen and wait at least 15 to 20 minutes before checking the P_aO_2. These patients have regions of the lung with long time constants (see Chapter 4 for a refresher on time constants), and it may take many minutes before the alveolar gas reflects the gas in the room as opposed to the gas that had been inspired with added oxygen.

Having calculated the A-aDO$_2$ on room air, determine if it is normal or abnormal, while taking into consideration that the normal range varies with age (the upper limit of normal is approximately 0.3 times the age of the person) (Quick Check 5-3). If the A-aDO$_2$ is normal, the hypoxemia must be attributable to a reduced PIO$_2$ or hypoventilation. All the other physiological causes of hypoxemia are characterized by an abnormally large A-aDO$_2$.

QUICK CHECK 5-3	NORMAL ALVEOLAR-ARTERIAL OXYGEN GRADIENT (BREATHING ROOM AIR)

AGE (yrs)	A-aDO$_2$ (mm Hg)
1–30	≤ 10
40	≤ 12
50	≤ 15
70	≤ 25

THOUGHT QUESTION 5-8: Two patients are sitting next to each other. Each has oxygen tubing that is connected to an oxygen tank. The patients are wearing the tubing on their faces such that oxygen is flowing into their noses. Each is receiving 2 L/min of 100% oxygen from the tanks. Do they have the same FIO$_2$? Explain your thoughts.

If the A-aDO$_2$ is abnormally large, place the patient on supplemental oxygen and recheck the P_aO_2. If there is a significant increase in the P_aO_2 (generally more than 20 mm Hg), you are likely dealing primarily with $\dot{V}/Q$ mismatch or a diffusion problem. If the P_aO_2 does not increase significantly, the patient probably has a shunt. (For very high levels of supplemental oxygen, the increase may be more than 20 mm Hg even with a fair degree of shunt. See Animated Figure 5-11 to appreciate the magnitude of change in P_aO_2 when giving 100% O$_2$ for different degrees of shunt.) Note that patients may have a combination of $\dot{V}/Q$ mismatch and shunt present at the same time; recall that shunt is actually the extreme form of $\dot{V}/Q$ mismatch when $\dot{V} = 0$. The absence of a change in the P_aO_2 with supplemental oxygen indicates that shunt is the primary cause of the hypoxemia. A large change in P_aO_2 with increased FIO$_2$ indicates that shunt is not the primary cause of the hypoxemia (Table 5-2).

If the patient has a normal P_aO_2 at rest but develops hypoxemia with exercise, one is likely dealing with a diffusion abnormality. One must remember, however, that more than one physiological derangement may be present in a given patient. For example, an individual with emphysema may have both $\dot{V}/Q$ mismatch and a diffusion abnormality.

TABLE 5-2 Physiological Causes of Hypoxemia

	A-aDO$_2$	SIGNIFICANT RESPONSE TO SUPPLEMENTAL O$_2$	PRESENT AT REST
Decreased PIO$_2$	Normal	Yes	Yes
Alveolar hypoventilation	Normal	Yes	Yes
Ventilation-perfusion mismatch	Increased	Yes	Yes
Shunt	Increased	No	Yes
Diffusion abnormality	Increased	Yes	Not usually, unless severe

PUTTING IT TOGETHER

You are working the night shift in the emergency department, and an unconscious man is brought in by the city's emergency medical services ambulance. You are asked to evaluate him. He is disheveled and smells strongly of alcohol. There is evidence of dried vomit on his shirt. He is breathing shallowly with at a rate of 6 breaths/min. His driver's license indicates that he is 60 years old. A pulse oximeter (a device that is put on the finger and quickly measures the percentage of oxyhemoglobin in the blood) indicates that the hemoglobin is only 80% saturated (normal is 95% to 97% saturated). The senior physician asks you if you think he aspirated his vomit and now has pneumonia to account for the low oxygen saturation. Before you answer her question, you indicate that you need some more data.

An arterial blood gas (ABG) is obtained with the patient breathing room air. The ABG shows $P_aO_2 = 54$ mm Hg; $P_aCO_2 = 65$ mm Hg; and pH = 7.20. Before reading on, calculate the P_AO_2 and the alveolar–arterial oxygen gradient; do you think there is a problem with this patient's gas exchanger?

Using the alveolar gas equation, you calculate the P_AO_2 and determine that it is 69 mm Hg. Thus, the alveolar–arterial oxygen gradient (A-aDO$_2$) is 15 mm Hg, which is within the normal range for a 60-year-old man. Thus, the cause of the hypoxemia must be either a decreased PIO$_2$ or alveolar hypoventilation. Because you are working at a hospital located at sea level and there is no history that the patient was in an environment with fire or smoke, a decreased PIO$_2$ seems unlikely. Furthermore, the P_aCO_2 is elevated, which is consistent with alveolar hypoventilation, in this case because of a probable reduction in total ventilation.

You give the patient supplemental oxygen with a mask that provides approximately 28% oxygen. You recheck the P_aO_2, which is now 120 mm Hg. The patient is beginning to wake up, and a serum alcohol level shows that he was quite intoxicated when he was brought into the emergency department. You tell the senior physician that you do not think the patient has pneumonia based on your analysis of the physiological cause of his hypoxemia. A chest radiograph confirms your impression.

Summary Points

- Alveolar ventilation is the amount of air that enters or exits the alveolus each minute and is one of the key determinants of the P_aCO_2.
- Dead space refers to regions of the lung that receive air but do not participate in gas exchange.

- Physiologic dead space is the sum of anatomic dead space (airways) and alveolar dead space (alveoli that receive air but are not perfused).
- Total ventilation is the sum of alveolar and dead space ventilation.
- Physiological dead space can be measured with determination of the dead space to tidal volume ratio using the Bohr method, which makes use of the principle that gas coming from dead space contains no carbon dioxide, while gas coming from perfused alveoli has a PCO_2 equivalent to the P_aCO_2.
- The carbon dioxide elimination relationship, or clearance equation for the lung, indicates that alveolar PCO_2 is inversely proportional to arterial PCO_2 and is directly proportional to carbon dioxide production.
- At FRC in an upright person, the alveoli at the apex of the lung are larger than at the base because of the distribution of the pleural pressure over the lung; the pleural pressure is more negative at the apex of the lung.
- During a normal tidal volume breath from FRC, the bases of the lung receive a greater proportion of the breath than the apices of the lung.
- The pulmonary circulation is a high-compliance, low-resistance system and can accommodate significant increases in blood flow with relatively small changes in pressure.
- In a normal person in the upright position, more blood flow goes to the bases than the apices of the lungs as a result of the effects of gravity.
- In the most superior (opposite gravity) portions of the lung, there may be no blood flow to some alveoli (alveolar dead space; zone 1 of the lung). In these regions of the lung, alveolar pressure may exceed pulmonary capillary pressure.
- The three zones of the lung describe the relative amounts of perfusion to different regions of the lung. These zones are derived from the relationships between alveolar, pulmonary arterial, and pulmonary venous pressures.
- The pulmonary arterioles respond to local hypoxia by vasoconstricting. This results in the redirection of blood flow to lung units with higher PO_2.
- Optimal efficiency of gas exchange depends on matching of ventilation and perfusion within the lung. $\dot{V}/Q$ mismatch is a major cause of hypoxemia in patients with cardiopulmonary diseases.
- The carbon dioxide–hemoglobin dissociation curve is relatively linear. This relationship allows hyperventilation of normal alveoli to compensate for hypoventilation of diseased lung units.
- The Haldane effect describes the shift to the right of the carbon dioxide–hemoglobin dissociation curve in the presence of oxygen. Carbon dioxide is displaced from hemoglobin and enters the blood as a dissolved gas. Thus, for a given CO_2 content, PCO_2 is higher.
- The physiological causes of hypercapnia include reduced alveolar ventilation (because of a decrease in total ventilation, a change in breathing pattern, or $\dot{V}/Q$ mismatch) and increased carbon dioxide production in the setting of minimal ventilatory reserve.
- The oxygen–hemoglobin dissociation curve has a sigmoid shape. This relationship ensures that a mild decrease in alveolar PO_2 does not significantly affect hemoglobin saturation and the oxygen content of the blood and that oxygen is readily released when blood reaches peripheral tissue.
- The majority of oxygen carried in blood is bound to hemoglobin.
- Anemia leads to reduced oxygen content in the blood but does not affect the P_aO_2.
- The oxygen-hemoglobin dissociation curve may shift to the right or the left based on body temperature, blood pH, and P_aCO_2.
- The alveolar gas equation allows one to calculate the P_AO_2. Using this value and the P_aO_2 obtained from an ABG, one can calculate the A-aDO_2, a number that is essential in analyzing the physiological cause of hypoxemia. An abnormal A-aDO_2 indicates that there is a problem with the gas exchanger.

- The five physiological causes of hypoxemia are decreased PIO_2, alveolar hypoventilation, $\dot{V}/Q$ mismatch, shunt, and abnormal diffusion. The physiological cause of hypoxemia can often be determined with knowledge of the A-aDO_2 and the response of the P_aO_2 to supplemental oxygen and with information on whether the hypoxemia is present at rest or only with exercise.
- $\dot{V}/Q$ mismatch is the most common physiological cause of hypoxemia. Mild to moderate $\dot{V}/Q$ mismatch is much more likely to produce hypoxemia than hypercapnia because of the different shapes of the oxygen-hemoglobin and carbon dioxide-hemoglobin dissociation curves.

Answers TO THOUGHT QUESTIONS

5-1. At the end of exhalation, the anatomic dead space is filled with the last gas that came out of the alveoli. Recall that the lung volume at the end of exhalation, FRC, is significant; thus, the alveoli still contain several liters of gas. If you ask the person to breathe in 100% oxygen and then exhale while you measure the exhaled volume of gas and the partial pressure of nitrogen or oxygen within it, you will get some useful information. The first gas to exit the mouth will be the last gas inhaled, that is, the 100% oxygen that is in the anatomic dead space. If you are measuring the PO_2, you will notice a high level of oxygen that will begin to decrease when gas from the alveoli begins to appear. If you were measuring PN_2, you will see no nitrogen while the gas from the anatomic dead space exits the mouth but will begin to see it appear as gas from the alveolus becomes apparent. In either case, you are able to make an assessment of anatomic dead space. This method of measuring anatomic dead space is called the Fowler method.

5-2. For a patient with a significant ventilatory pump problem, an increase in alveolar ventilation requires a great deal of mechanical work, which usually produces a sense of respiratory discomfort, muscle fatigue, or both. If the patient's controller resets itself (by mechanisms that remain unknown to us) and the patient now lives with a higher P_aCO_2 (assuming the kidneys can compensate for the resulting respiratory acidosis; see Chapter 7), then a lower alveolar ventilation will be required to eliminate the CO_2 being produced each minute. Thus, chronic hypercapnia, or elevations in arterial PCO_2, can be viewed teleologically as a way to reduce the work of breathing and respiratory discomfort.

5-3. With the institution of a diet rich in carbohydrates, carbon dioxide production will increase. If the respiratory system is unable to compensate by increasing alveolar ventilation, then the P_aCO_2 must increase. There are case reports of such a change in diet leading to respiratory failure in patients with severe chronic obstructive pulmonary disease. Such patients typically are fed diets relatively low in carbohydrates to diminish the need for high alveolar ventilation and thus reduce the work of breathing that must be done in the setting of an impaired ventilatory pump.

5-4. During exercise, cardiac output increases, which leads to an increase in blood flow to the lungs. As flow increases, more blood goes not only to the bases of the lungs, but one also begins to recruit vessels in the upper zones that received little, if any, blood flow under conditions of rest. This process reduces the amount of dead space in the lung (recall that there is relatively more ventilation than perfusion to the apices of the lung) and improves $\dot{V}/Q$ matching. In patients with emphysema, in whom there is usually considerable abnormality in the airways of the upper portions of the lung, exercise may worsen $\dot{V}/Q$ matching by sending blood to areas with increased dead space. In such patients, V_D/V_T may increase with exercise. Measurement of V_D/V_T in these circumstances can provide clues to the presence of otherwise unsuspected lung disease.

5-5. In outer space, under conditions of weightlessness, the effects of gravity on the distribution of ventilation and perfusion in the lung are negated. Thus, one might expect there to be a more even distribution of both ventilation and perfusion throughout the lungs. However, data from animal studies suggest this effect is not as great as expected if gravity were the sole determinant of distribution of blood flow in the lungs. Regional differences in the anatomy of the vasculature may also contribute to the distribution of flow.

5-6. All of the factors that cause the hemoglobin–oxygen dissociation curve to shift to the right either cause or are the consequence of an increase in oxygen consumption within the tissues (increased body temperature, increased P_aCO_2, decreased pH). The shift of the curve to the right results in greater release of oxygen to the tissues. Thus, the evolutionary advantage of the shift is to try to restore the organism to normal conditions.

5-7. The blood in the right heart has just returned from the extremities and central organs and contains relatively little oxygen. If it now moves to the left side of the heart and is pumped from there back out to the body without first traveling to the lungs to pick up oxygen, the P_aO_2 will be reduced. Giving this patient supplemental oxygen will not have a significant effect on the P_aO_2 because the blood going through the atrial septal defect never perfuses the lung and will not see the higher FIO_2. As noted with a pulmonary shunt, the blood perfusing normal alveoli is already nearly fully saturated, and the increased FIO_2 will have little effect on the oxygen content of that blood. This condition is called an intracardiac shunt.

5-8. It is most likely that the patients have different FIO_2. Although they are each receiving 2 L/min of oxygen from the tanks, the concentration of oxygen that ultimately enters the lungs is determined by the amount of gas being inhaled with the oxygen. The gas inhaled from the room dilutes the oxygen coming from the tubing. Remember, normal minute ventilation is approximately 5 L/min. Thus, even at normal levels of ventilation, the oxygen is being mixed with several liters of gas that has only 21% oxygen. If the two patients have different minute ventilation, which is likely, they will have different FIO_2.

Review Questions

DIRECTIONS: Each of the numbered items or incomplete statements in this section is followed by answers or by completions of the statement. Select the ONE lettered answer or completion that is BEST in each case.

1. Complete the missing elements in the table below that summarizes the factors used to analyze the physiological causes of hypoxia.

PHYSIOLOGICAL CAUSE OF HYPOXIA	A-aDO$_2$ (NORMAL OR INCREASED)	SIGNIFICANT RESPONSE TO SUPPLEMENTAL O$_2$ (Y/N)	PRESENT AT REST (Y/N)
Decreased PIO$_2$			
Alveolar hypoventilation			
Ventilation-perfusion mismatch			
Shunt			
Diffusion abnormality			

Questions 2, 3, 4 are based on the following case.

You are taking care of a patient in the intensive care unit (ICU). She is 40 years old and has only one lung (she underwent a left pneumonectomy several years earlier because of recurrent infections), and that lung is damaged by severe scarring or pulmonary fibrosis. She entered the ICU with respiratory failure caused by severe bronchitis. She is very weak, and you are having trouble getting her to sustain her own breathing without the ventilator. On the ventilator with an FIO$_2$ of 0.35, her arterial blood gas shows P_aO_2 = 80 mm Hg; P_aCO_2 = 60 mm Hg; and pH = 7.34 with a respiratory rate of 25 and a tidal volume of 350 mL. When you take her off the ventilator, she immediately increases her respiratory rate to 40 breaths/min with a tidal volume of 250 mL. Assume the patient has an anatomic dead space of 125 mL and an alveolar dead space of 50 mL.

2. The alveolar–arterial oxygen gradient is approximately:

A. 95 mm Hg
B. 125 mm Hg
C. 175 mm Hg
D. 225 mm Hg
E. 250 mm Hg

3. The physiological cause of the patient's hypoxemia is:

A. decreased PIO_2
B. alveolar hypoventilation
C. $\dot{V}/Q$ mismatch
D. shunt
E. diffusion abnormality
F. B and C
G. D and E

4. To increase the chance that the patient will be able breathe on her own without support from the ventilator, you should prepare her by doing the following with the ventilator:

A. Change the ventilator settings so that the P_aCO_2 will be 40 mm Hg within 2 days.
B. Leave the ventilator settings as they are now, with a resulting P_aCO_2 of 60 mm Hg.
C. Change the ventilator settings so that the P_aCO_2 will be 80 mm Hg within 2 days.

5. You are working in the emergency department, and a patient comes in with a complaint of shortness of breath and chest pain. There is a strong clinical suspicion that he may have a pulmonary embolism (i.e., a blood clot that travels from a vein, usually in the legs, to the lungs). You recall that most patients with pulmonary embolism have hypoxemia. You ask the nurse to check the patient's oxygen saturation using a pulse oximeter (an instrument that is placed on the finger and assesses the oxygen saturation of hemoglobin passing through capillaries in the nail bed), and she tells you that it is normal–the oxygen saturation is 96%. With these results you should:

A. Be reassured that the patient does not have pulmonary embolism and pursue other diagnoses.
B. Request an arterial blood gas.
C. Give supplemental oxygen to determine if the saturation improves.

The Controller: Directing the Orchestra

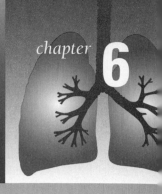

CHAPTER OUTLINE

LEARNING OBJECTIVES

- **To enumerate the many factors that contribute to the control of breathing.**
- **To establish the dual nature (voluntary and automatic) of the control of the respiratory system.**
- **To define the anatomy and function of the receptors, the stimulation of which affect the rate and depth of breathing.**
- **To delineate the ventilatory responses to hypoxemia and hypercapnia.**
- **To describe the impact on the controller of supplemental oxygen in patients with acute on chronic respiratory failure.**

At this moment you are, hopefully, sitting in a comfortable chair, reading this book. Every 5 seconds or so, you take a breath. Why? Most people respond to that question by answering, "Because we all need oxygen to live." Although that statement is true, consider a second question before deciding if a need for oxygen really explains why we breathe. How low does your P_aO_2 decrease in the 5 seconds between breaths? The answer to that is, "Not very much," perhaps not even a measurable amount. So, if we do not breathe in response to hypoxemia, you might think it is because we are becoming hypercapnic; our P_aCO_2 must be increasing. Again, we ask you to consider another question, "How high does your carbon dioxide increase in the 5 seconds between breaths?" As you probably have guessed by now, the answer to that question is also, "Not very much." So why do we breathe?

The physiological control of the respiratory system is unique among organ systems. Because breathing is essential to life and must occur 24 hours a day, 365 days a year, whether you are awake or asleep, conscious or unconscious, there must be an automatic mechanism to determine the rate and depth of breathing on a minute-by-minute basis. Most enzymes and chemical reactions in the body operate best within a very narrow range of oxygen level and pH. Even short periods of absent breathing or breathing in excess of the metabolic needs of the body can lead to life-threatening derangements in the internal milieu. In this sense, the respiratory system is similar to the cardiovascular, endocrine, and gastrointestinal systems, which are all regulated without the need for conscious instructions.

At the same time, humans (and other mammals, for that matter) need to be able to temporarily interrupt the normal pattern of breathing to perform other functions such as vocalizing, eating, and lifting heavy objects. Imagine, for example, that you are just about to swallow some food and your inspiratory muscles are activated, thereby creating a negative intrathoracic pressure. Material in the posterior portion of the pharynx could be sucked down into the trachea and lungs, leading to obstruction of the airways and pneumonia. To prevent such mishaps and to use ventilatory muscles for other functions at times, we possess the ability to temporarily change our breathing patterns. To call to someone down the block, you can instantly take a deep breath and forcefully exhale to shout a loud "stop!" Alternatively, to swim under water, you can hold your breath, at least for a while, to avoid aspiration. This voluntary control of the respiratory system is unique—you certainly cannot make your heart double its rate or stop completely by willing it to happen, nor can you control peristalsis or consciously instruct your adrenal gland to secrete epinephrine.

In this chapter we explore the role of the cerebral cortex (the volitional component to the controller) and the brainstem (the automatic component) in breathing. In addition, we describe many receptors that, when stimulated, provide information to the brain and modify the output from both the cortex and the brainstem (Fig. 6-1). The chapter contains a moderate amount of detail about the receptors in the upper airways, lungs, and chest wall,

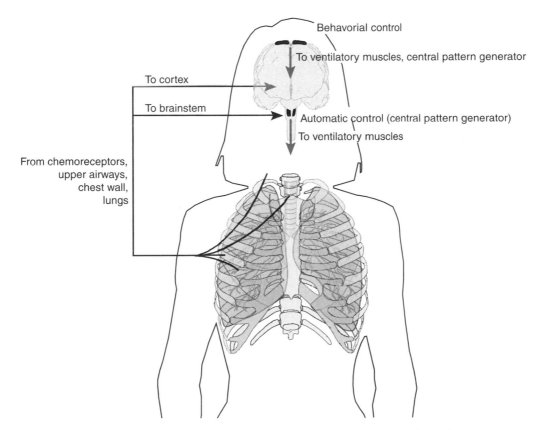

FIGURE 6-1 The elements of ventilatory control. Multiple factors contribute to the control of breathing. Conscious or volitional control arises from the cerebral cortex, and automatic control emanates from the brainstem. Information from mechanical and chemoreceptors throughout the respiratory system modulates the activity of the control centers.

not only because of their relevance for the physiology of ventilatory control but also because an understanding of these receptors is essential for our exploration of respiratory sensations and the symptom of "shortness of breath" that are pursued in Chapter 8.

The control of breathing is a very complex process, especially when the heart and lungs become diseased. Nevertheless, there are doctors you will encounter in the next few years who will tell you it is all about low oxygen or high carbon dioxide. We are confident that by the time you finish this chapter, you will understand and appreciate the beauty and complexity of the ventilatory controller.

Definitions

As we begin our exploration of the control of ventilation, it is important to be sure we are all using terms in the same way. Some of these words were familiar to you before you embarked on your medical studies, but the meanings you associated with them at that time may not be the same as the ones we use in the care of patients.

Hyperventilation is a term that describes breathing beyond that which is required to meet the metabolic needs of the body as reflected in the production of carbon dioxide. Recall the equation (discussed in Chapter 5) that relates P_aCO_2 to carbon dioxide production and alveolar ventilation:

$$\dot{V}_A = K \times \frac{\dot{V}_{CO_2}}{P_aCO_2}$$

When alveolar ventilation ($\dot{V}_A$) is well matched to carbon dioxide production ($\dot{V}_{CO_2}$), P_aCO_2 is normal. Thus, if a person is hyperventilating, it means that the P_aCO_2 is lower than the normal range, that is, below approximately 38 mm Hg. On the other hand, **hypoventilation** refers to breathing that is less than required to meet the metabolic needs of the body, and it is characterized by a P_aCO_2 that is above the normal range, that is, greater than approximately 42 mm Hg.

Hyperpnea is increased breathing that matches the metabolic needs of the body, again as reflected in the production of carbon dioxide. If ventilation is increasing in concert with carbon dioxide production, P_aCO_2 remains within the normal range. During most of exercise, for example, ventilation increases in proportion to metabolic needs, and you would refer to the increased breathing as hyperpnea.

When a person increases ventilation, one of two strategies may be used: the respiratory rate may be increased, or the tidal volume (V_T) may be increased (of course, you can do both at the same time). **Tachypnea** refers to an increase in respiratory rate above the normal range (this term is usually reserved for rates greater than or equal to 20 breaths per minute). You must remember, however, that a person can be tachypneic without hyperventilating. As discussed in Chapter 5, if rapid breathing is accompanied by small tidal volumes, the alveolar ventilation may be reduced, even if total ventilation is increased. This could result in an increase in P_aCO_2, and you would describe the patient as being tachypneic and yet showing evidence of alveolar hypoventilation. Finally, **bradypnea** is a term used to describe respiratory rates below the normal range (this term is usually reserved for rates less than 10 breaths per minute).

Central Control

THE AUTOMATIC CENTERS

The neural structures responsible for the automatic control of breathing appear to be in the medulla. Two aggregates of neurons, termed the dorsal respiratory group (DRG) and the ventrolateral group (VRG), have both inspiratory and expiratory neurons. In addition, the DRG seems to play an important role in processing information from receptors in the lungs, chest wall, and chemoreceptors that modulate breathing. Neural activity from the DRG is important in activation of the diaphragm, and the VRG, in addition to having a role in determining the rhythm of breathing, regulates the changes in diameter of the upper airway that occur with breathing by stimulating muscles to expand the upper airway during inspiration.

THOUGHT QUESTION 6-1: What would happen to inspiratory flow if the neurons in the VRG were damaged by a stroke? Why?

There are also neurons in the pons (the pontine respiratory group [PRG]) that may contribute to the transitions or switching from inspiration to expiration. Damage to the respiratory neurons in the pons leads to an increase in inspiratory time, a decrease in respiratory frequency, and an increase in the tidal volume.

Within the medulla, there appear to be inspiratory neurons that have a pacemaker function, that is, they have intrinsic properties that give them rhythmic activity. Similar to the cardiac pacemaker, they fire at a particular rate that can be modified by other factors, such as disturbances in gas exchange or stimulation of pulmonary receptors, which are discussed at greater length later in this chapter. Information from pulmonary receptors is transmitted to and processed in the DRG. This feedback from the lungs to the medulla is believed to affect the respiratory pattern.

Some of the neurons in the medulla fire during inspiration; others seem to have a role in the transition from inspiration to expiration; and others appear to fire during expiration, primarily serving an inhibitory role on the diaphragm. Neurons in the DRG increase their rate of firing during inspiration. Others in the VRG seem to increase activity during expiration. Taken together, all of these respiratory neurons responsible for our automatic rhythmic breathing are termed the **central pattern generator (CPG)**.

If you examine the activity of the inspiratory neurons during a respiratory cycle, you see three phases of activity (Fig. 6-2). During the inspiratory phase of the respiratory cycle, the

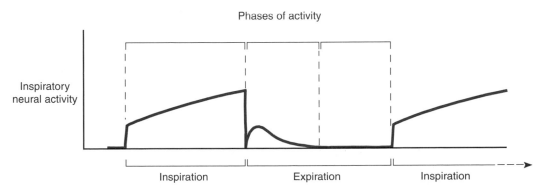

FIGURE 6-2 Inspiratory neural activity during the respiratory cycle. During inspiration, the activity of the inspiratory neurons rapidly increases. Inspiration is terminated by the activity of inhibitory neurons. But there appears to be a second activation of inspiratory activity at the beginning of expiration that may serve to slow expiratory flow and smooth expiration (remember that expiration is "passive" during normal breathing). (Adapted from Leff A, Schumacker P: *Respiratory Physiology: Basics and Applications,* 1st ed. Philadelphia: WB Saunders, 1993, p. 116.)

activity of the inspiratory neurons rapidly increases. This phase appears to be terminated or switched off by the action of inhibitory neurons. At the beginning of expiration, however, there may be another burst of inspiratory activity that serves to slow or put a brake on expiratory flow. In asthma, this inspiratory activity during expiration appears to be accentuated and may contribute to the hyperinflation seen in individuals with that disease.

THOUGHT QUESTION 6-2: Hold your breath while keeping your glottis open. How did you do that? Are you using expiratory muscles to keep you from inhaling? What is happening at the level of the central pattern generator?

VOLITIONAL BREATHING

If someone asks you to hold your breath, you can for a short time. If someone asks you to breathe at 30 breaths a minute, you can respond appropriately. If someone asks you to take a breath with a tidal volume of 2 L, you inhale deeply. Clearly, you can send messages from your motor cortex to your ventilatory muscles. These neural impulses descend in the spinal cord in corticospinal tracts that are separate from the tracts containing axons originating in the automatic centers. The nerves that innervate the diaphragm exit the spinal cord in the midcervical level (C3–C5). Thus, people who sustain spinal cord injuries below this level in the cervical cord may be severely impaired with quadriplegia, but they can still inspire normally. The intercostal muscles are innervated by nerves exiting the spinal cord along the thoracic column, so they may not function, depending on the level of a spinal cord injury. But as long as the diaphragmatic innervation is intact, inspiration can still take place. Ventilation, therefore, is much more severely compromised by a cervical spinal cord lesion, which prevents activation of both the diaphragm and intercostal muscles, than by a thoracic spinal cord injury.

When you are awake, there may be elements of volitional control of breathing, even when you are not consciously thinking about taking a breath. When asleep, a person's P_aCO_2 tends to be higher than when awake, for example. An unusual medical problem, called congenital central hypoventilation syndrome (CCHS), offers another example of "volitional" control in the absence of actual thought directed to breathing. CCHS is characterized by the absence of a functioning central pattern generator. In this rare disorder (fewer than 200 cases are reported in the medical literature), infants stop breathing when they go to sleep, and they are identified shortly after birth when they suffer a respiratory arrest during sleep. These individuals must be maintained on ventilators at night throughout their lives. When awake, however, they breathe quite normally, although their respiratory pattern is a bit more irregular than those without this syndrome. The presumption is that the reticular activating system may have connections to the respiratory control areas in the brain that help to regulate breathing when we are awake.

The concept of volitional control of breathing also includes the effects of discomfort and anxiety on our breathing. When experiencing pain or shortness of breath, most people increase their respiratory rate, and total ventilation increases. The pattern of breathing may also reflect attempts to reduce the discomfort associated with ventilation. Patients with significantly reduced respiratory system compliance, for example, tend to breathe with a rapid, shallow pattern. Because the system is stiff, it requires less work to breathe in this way than to exert the large intrathoracic pressures necessary to distend the system. For patients with increased airway resistance, on the other hand, the high flow required for rapid, shallow breathing requires considerable work. These patients tend to adopt a slower breathing pattern with large tidal volumes. Anxious patients typically increase their respiratory rates.

SOURCES OF INFORMATION THAT MODULATE THE CONTROL OF BREATHING

The central control centers for the respiratory system do not act in isolation. The brain constantly receives information from the upper airways, lungs, and chest wall that tell it how the ventilatory pump is responding to the messages exiting the controller. Most of the receptors in the upper airways, lungs, and chest wall are **mechanoreceptors**, so called because they are activated by mechanical distortion of their local environment. In addition, the status of the gas exchanger is also followed closely, and the controller reacts to decreases in the P_aO_2 and increases in P_aCO_2. Information arising from the periphery (the so-called afferent neural pathways) travels to the brain. For example, sensory input from the upper airways, lungs, and peripheral chemoreceptors travels up the ninth and tenth cranial nerves to the region in the medulla where the DRG is located. We will now examine some of the key sources of afferent information that affect the activity of the central controller.

Upper Airway Receptors

Some receptors in the airways appear to sense and monitor flow. These receptors likely respond, however, to changes in temperature resulting from the flow of air past them rather than directly to flow itself. Stimulation of flow receptors appears to inhibit the central controller.

Contained within the walls of the pharynx are receptors that appear to be activated in association with swallowing. Respiratory activity ceases during swallowing as the epiglottis covers the larynx. From an evolutionary standpoint, this sequence of events minimizes the risk of aspiration of food and liquid into the lungs.

THOUGHT QUESTION 6-3: A patient with chronic obstructive pulmonary disease (COPD) comes to see you for increased shortness of breath. It is a hot July day. The temperature is 100°F, and the humidity is 95%. The patient says, "It just doesn't feel like there is any air going in." You assess his lung function and arterial blood gas. Nothing has changed from the last time you saw him 3 months ago. Why might he be feeling the way he does?

Pulmonary Receptors

The lungs contain two types of stretch receptors, both of which are myelinated. In addition, C fiber endings may monitor changes in the pulmonary circulation and interstitial space and transmit information to the brain via unmyelinated fibers in the vagus nerve.

Stretch receptors in the lungs are categorized based on the speed with which the rate of firing of the receptor changes as the local environment, specifically, changes in lung volume, in which the receptor is located is altered. **Slowly adapting stretch receptors (SARs)** continue to fire at a fairly constant rate after being stimulated by a stretch, even after they have reached a new constant length (i.e., they adapt slowly to the new lung volume). In contrast, **rapidly adapting stretch receptors (RARs)** change their firing rate quickly upon reaching a new level in the environment (Fig. 6-3).

Use Animated Figure 6-3 to observe how the SARs and the RARs respond to a change in lung volume. Listen as you watch the neural impulses to get a feel for how quickly each type of receptor adapts to the new conditions.

It is believed that the SARs are located among smooth muscle cells within the intra- and extrathoracic airways. When these airway receptors are stimulated by lung inflation, the expiratory phase of respiration is prolonged. These receptors may also play a role in the early termination of inspiration when tidal volume increases. Deflation of the

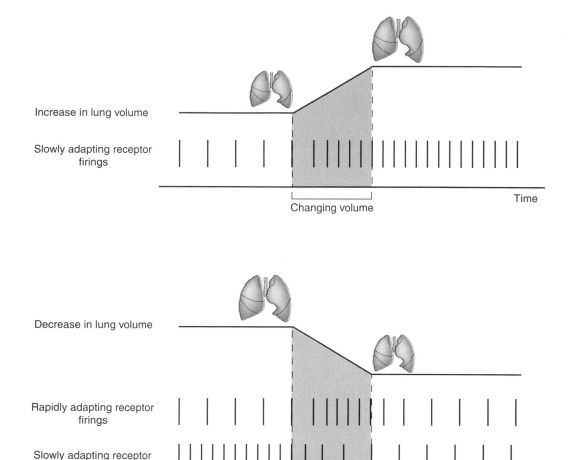

FIGURE 6-3 Slowly and rapidly adapting stretch receptors (SARs and RARs). SARs increase their firing rate when stimulated by inflation and decrease their firing rate on deflation, and they continue to exhibit an altered firing rate, even after a new lung volume is achieved (because the rate of firing changes slowly after the stimulus is released, the receptors are "slowly adapting" to the new state). The activation of these receptors on inflation helps to "turn off" inspiration. In contrast, the rate of firing of the RARs, which are stimulated during deflation, quickly returns to baseline after a new lung volume is achieved (the rate of firing changes quickly after the stimulus is release, and the receptors are "rapidly adapting" to the new state). Activation of these receptors is believed to send a signal to the controller to begin the next inhalation and may lead to an increase in respiratory rate as well as to the deep breaths that we call "sighs." (The effect of RARs on the controller during inflation is unclear and may vary with lung volume.) **Also see Animated Figure 6-3.**

lung causes a decrease in the basal firing rate of the SARs, which appears to lead to an increase in the respiratory rate.

The RARs are thought to be located within the airways in association with airway epithelial cells, with the majority appearing to be near the region of the carina and in the larger bronchi. The RARs can be stimulated both by chemical (e.g., cigarette smoke, histamine, prostaglandins) and mechanical stimuli. An older name for some of these nerves is *irritant receptors*, which reflects their activation in the presence of chemical stimuli that are perceived as irritating to the lungs. Activation of the more centrally located RARs may lead to cough, bronchospasm, and increased mucus production, and stimulation of

TABLE 6-1 Airway and Pulmonary Receptors			
LOCATION	TYPE	STIMULUS	EFFECT ON VENTILATORY CONTROL
UPPER AIRWAY			
Nose	Mechanical	Flow	Decrease ventilation
Pharynx	Mechanical	Swallow	Stop breathing
PULMONARY			
SARs	Mechanical	Lung inflation	Prolong expiratory time Terminate inspiration
		Lung deflation	Increase respiratory rate
RARs	Mechanical and chemical	Lung deflation	Increase respiratory rate Sighs
C fibers	Chemical and mechanical	Increased pulmonary capillary pressure, lung inflation, chemicals such as histamine, bradykinin, prostaglandins	?Increased respiratory rate, reduced tidal volume

RARs = rapidly adapting stretch receptors; SARs = slowly adapting stretch receptors.

the receptors located more deeply within the lungs can lead to hyperpnea. Lung deflation activates these receptors and can contribute to an increase in respiratory rate as well as the periodic large breaths that we take (i.e., sighs).

The unmyelinated fibers, termed **C fibers**, in the lungs carry information from a variety of receptors whose function is not totally understood. A group of C fibers that arise from deep within the lungs are believed to carry information from **J receptors**. These receptors are located near (or in juxtaposition with, hence the J in J receptors) to the pulmonary capillaries. Researchers believe that these receptors may be activated by increases in pulmonary capillary pressures or the accumulation of interstitial fluid, as is seen in the setting of congestive heart failure (CHF). C fibers may also arise from receptors in bronchi. Both chemical and mechanical factors may stimulate receptors that are served by C fibers, some of which may play a role in bronchoconstriction. The exact role of C fibers in the modulation of respiratory control is uncertain, but it is possible that they contribute to the tachypnea seen in patients with CHF and other conditions that are associated with acute changes in pulmonary capillary pressure (Table 6-1).

Chest Wall Receptors

The primary receptors in the chest wall that have a role in monitoring respiration are the muscle spindles and Golgi tendon organs. Their function appears to pertain primarily to alerting the controller to the fact that the physiology of the ventilatory pump has changed, that is, that airway resistance has increased or respiratory system compliance has decreased. Such changes are generally characterized as an increased "load" on the respiratory system. In other words, for a given **efferent**, or outgoing neural discharge from the brain to the muscles, the respiratory system is not responding appropriately. Having received **afferent**, or incoming signals from these peripheral receptors to the brain, the typical response of the controller to an increased load is to increase the efferent activity to the muscles in an effort to "compensate" for the load.

Muscle spindles are located in the skeletal muscles. They are found in intercostal muscles and, to a lesser degree, in the diaphragm. The muscle spindle responds to mechanical stimuli (contraction of the muscle) and behaves like a slowly adapting receptor. The spindle receptor is attached to spindle muscle fibers that are innervated by gamma fibers. The spindle muscles are arranged in parallel with the main contracting fibers of the muscle. When a message is sent from the brain to the ventilatory muscle to contract, a message sent over alpha motor neurons, a simultaneous message is sent to the spindle muscle over the gamma fibers. When the spindle muscle contracts, the spindle receptor is stretched and activated. If the main muscle, which is simultaneously receiving a message to contract, also contracts *and shortens*, the effect on the spindle is to reduce its tension to baseline, and there will be no change in output from the spindle receptor (Fig. 6-4).

Thus, the spindle activity provides feedback on the response of the chest wall to the command to the ventilatory muscles to contract. If there is no load on the respiratory system, the ventilatory muscles will shorten significantly, the spindle will no longer be stretched, and the firing frequency will be diminished. If there is a high mechanical load on the ventilatory pump (imagine the extreme case in which a person has aspirated a piece of food and the trachea is nearly completely occluded), the ventilatory muscles will

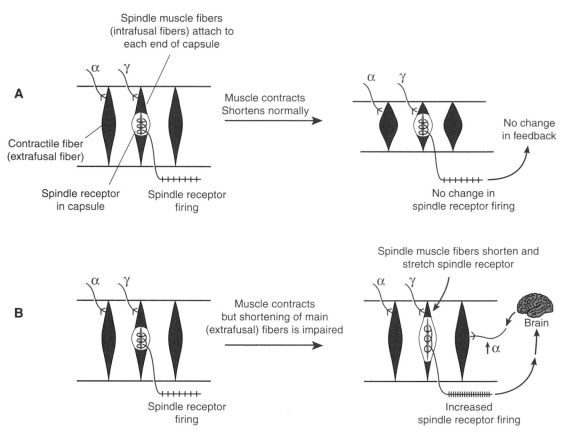

FIGURE 6-4 Muscle spindles and load compensation. The main contractile fibers of the ventilatory muscle are innervated by an alpha fiber. The muscle spindle is arranged in parallel to the main contractile fibers and contains muscle fibers that are attached to the sensory apparatus within the spindle. **A,** Both the contractile fiber and the muscle fiber within the spindle are activated simultaneously during inspiration. The spindle receptor is activated by lengthening in response to the contraction of the muscle fibers within the spindle and is deactivated by shortening that occurs when the main muscle fiber shortens. The net result is no change in output from the spindle receptor. **B,** If there is a mechanical load on the ventilatory pump, the main muscle fibers shortens very little. The contraction of the spindle muscle, however, lengthens and activates the spindle receptor, which remains activated and sends messages to the brain, resulting in greater output to the main muscle fibers. Thus, the system has tried to compensate for the load on the ventilatory pump. **Also see Animated Figure 6-4.**

contract but will shorten minimally because little air can enter the lung because of the region of high resistance in the trachea. The spindle muscles, however, will shorten, the spindle receptor will be activated and will remain activated, and the receptor will send afferent messages to the brain to further increase the contraction of the ventilatory muscles. As a result, **load compensation** will have occurred, that is, the controller will have made an adjustment to the presence of a severe resistive load on the ventilatory pump by instructing the muscles to contract more forcefully.

 Use Animated Figure 6-4 to explore the role of muscle spindles in load compensation. Observe the activity of the spindle receptors when they respond to different loading conditions. Note that for conditions of high load with poor muscle shortening (as with the aspirated bolus of food blocking airflow, for example), the feedback from the muscle spindle leads to a further increase in the contraction force of the muscle.

Golgi tendon organs are located at the point of insertion of the muscle fiber into the tendon. They appear to monitor tension on the muscles, and it is thought that when tension is very high, messages from these receptors inhibit further contraction. This mechanism may serve to protect muscles from injury when the load on the system is very high.

Chemoreceptors

The controller monitors changes in P_aO_2, P_aCO_2, and arterial pH via specialized neural structures called **chemoreceptors**. The **peripheral chemoreceptors**, located in the carotid bodies and the aortic arch, respond to changes in P_aO_2, P_aCO_2, and arterial pH. The **central chemoreceptors**, located in the medulla, respond to alterations in PCO_2 and pH in the cerebrospinal fluid (CSF), which reflects changes in these variables in the arterial blood.

The carotid bodies are found at the bifurcation of the common carotid artery into the internal and external carotid arteries. Because the chemoreceptors receive more blood than they need to meet their local metabolic needs, the P_aO_2 within the chemoreceptor reflects the delivery and consumption of oxygen by the rest of the body. This allows the chemoreceptor to sense and respond to the requirements of the body as a whole. The capillaries that perfuse the chemoreceptors are intimately associated with specialized nerve endings. Information from these nerve endings is carried to the brain via the ninth cranial nerve. It is not entirely clear how the carotid bodies sense hypoxemia, but it is clear that the stimulus for increased ventilation is P_aO_2, not the oxygen content of the blood. At normal levels of P_aO_2, some neural activity arises from the carotid bodies. At hyperoxic (above normal) levels, this activity is reduced but does not cease. As P_aO_2 decreases below 60 mm Hg, the rate of firing rapidly increases. Also, in most mammals, some tissues in the aortic body, located in the arch of the aorta, appear to be sensitive to hypoxemia. Removal of the carotid bodies in humans, however, abolishes the ventilatory response to hypoxemia, which suggests that the aortic bodies play little role in ventilatory control in people.

THOUGHT QUESTION 6-4: Does a person with anemia have an increased ventilation at rest? Why or why not?

The activity of the peripheral chemoreceptors also increases with high levels of P_aCO_2 (leading to increased ventilation) and reduced levels of arterial pH (leading to increased ventilation). Because elevations of carbon dioxide in the blood are also associated with a decrease in pH (more in Chapter 7), it is not immediately evident whether P_aCO_2 or pH is the stimulus under conditions of acute hypercapnia. Experiments in which the two variables are modified independently, however, suggest that the carotid body can respond to either stimulus. Although responsive to changes in both oxygen and carbon dioxide levels, the chemoreceptor is much more sensitive to acute hypercapnia than to hypoxemia. (Note: we may refer to the peripheral and central chemoreceptors in the singular, as above, even though each is composed of many nerve endings).

The central chemoreceptor appears to be a less discrete anatomic site than the peripheral chemoreceptors. Nerves that respond to change in PCO_2 (by increasing ventilation in response to increased PCO_2) and pH (by increasing ventilation in response to a decreased pH) appear to be located in both ventral and dorsal regions of the medulla. The ventral regions, in particular, appear to be near the surface of the brainstem in close proximity to CSF. This location may facilitate the ability of the central chemoreceptor to monitor changes in PCO_2 and pH. Changes in arterial PCO_2 and pH alter the levels of carbon dioxide and protons in the CSF, although, as we will discuss shortly, at different rates. As with the peripheral chemoreceptor, under experimental conditions, either an elevated PCO_2 or a reduced pH can independently cause an increase in respiratory-related neural activity from these regions. Our experience with mixed acid–base disorders, however, suggests that pH may be a more important factor in the regulation of breathing.

In assessing the response of the controller to changes in P_aO_2, P_aCO_2 and arterial pH, one must also examine the interactions of the peripheral and central chemoreceptors. Because of the blood–brain barrier, the neurons in each receptor may have somewhat different local environments at a given point in time. This can cause the chemoreceptors to seem to be out of phase with each other. In fact, it serves to smooth out the response to an acute change in gas exchange or the acid–base status of the body.

The blood–brain barrier has a differential permeability to ions such as H^+ (low permeability) and lipid-soluble molecules such as carbon dioxide (high permeability). If one were to infuse an acid into the blood, the peripheral chemoreceptor would respond by increasing ventilation before the local environment in the fluid bathing the medulla reflected the acid pH in the blood. As ventilation increases by virtue of the stimulation of the peripheral chemoreceptor, the P_aCO_2 decreases, which results in the diffusion of carbon dioxide from the fluid surrounding the brain back into the blood. The environment of the central chemoreceptor would rapidly reflect the lower PCO_2, but only later reflect the elevated H^+ concentration of the blood (because of the extra time needed for the H^+ ions to cross the blood-brain barrier, as already mentioned). The activity of the central chemoreceptor would decrease in the short term, which would attenuate the body's total response to the acid challenge. Alternatively, if one infused a buffer into the blood, such as sodium bicarbonate, and the arterial pH level increased, the activity of the peripheral chemoreceptor would decrease, ventilation would decrease, and the P_aCO_2 level would increase. Carbon dioxide would then diffuse across the blood–brain barrier and increase the PCO_2 level in the brain. Again, because the equilibration of H^+ across the blood–brain barrier occurs more slowly, in the short term, the activity of the central chemoreceptor would increase (Fig. 6-5).

 Use Animated Figure 6-5 to acutely change the pH and PCO_2 of the blood, and observe the changes in the fluid bathing the brain, as well as the effects on the activity of the peripheral and central chemoreceptors. In particular, notice how the central chemoreceptors can serve to attenuate the short-term ventilatory response to an acid or base load.

The responses described thus far reflect the sequence of events that occurs as the result of acute changes in carbon dioxide or pH. Both the blood and the brain have mechanisms to restore pH toward normal levels when acute disruptions occur. Thus, when P_aCO_2 is elevated chronically, as might occur in a patient with severe COPD, the activity of the peripheral and central chemoreceptors decrease within a few days as pH is normalized. At extremely high levels of carbon dioxide ($PCO_2 > 80$–100 mm Hg), an anesthetic effect may be produced, and ventilation decreases rather than increases (Table 6-2).

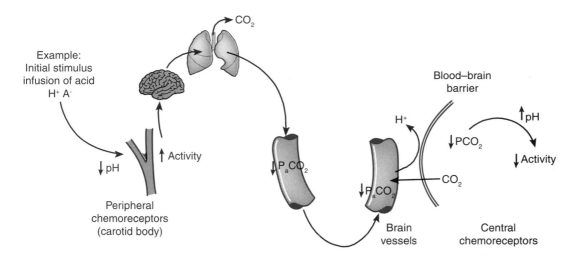

*Short-term response shown.

FIGURE 6-5 Interactions between peripheral and central chemoreceptors. Because of the blood–brain barrier and its relative impermeability to ions such as H^+ compared with fat-soluble molecules such as carbon dioxide, the local environments and resulting responses of the peripheral and central chemoreceptors may vary when the composition of the blood is acutely changed. **Also see Animated Figure 6-5.**

THE VENTILATORY RESPONSE TO HYPOXEMIA

One can characterize the ventilatory response of an individual to acute derangements in P_aO_2 and P_aCO_2 with a laboratory evaluation in which the person breathes a mixture of gases that alter the partial pressure of the gas in the blood in a progressive manner. The ventilation is measured continuously and plotted as a function of the partial pressure of oxygen or carbon dioxide (note: in healthy lungs, the alveolar and arterial PO_2 levels will

TABLE 6-2 The Response of the Controller to Acute and Chronic Changes			
STIMULUS	PERIPHERAL CHEMORECEPTORS	CENTRAL CHEMORECEPTORS	RESPONSE OF THE CONTROLLER
HYPOXEMIA			
Acute	↑↑	↓	↑
Chronic	↑↑↑	0	↑↑↑
HYPERCAPNIA			
Acute	↑↑	↑↑↑	↑↑↑↑↑
Chronic	↑	↑↑	↑↑↑
MILD METABOLIC ACIDOSIS			
Acute	↑↑	↓	↑
Chronic	↑	0	↑

Note: The arrows denote relative changes in activity of the chemoreceptors and for the controller, the change in the total ventilation. The hypoxemic condition is for someone at sea level. Other factors come into play in the response to chronic hypoxemia at altitude. Note that the decreased activity of the central chemoreceptors in response to hypoxemia is an indirect effect from the resulting hyperventilation and hypocapnia.

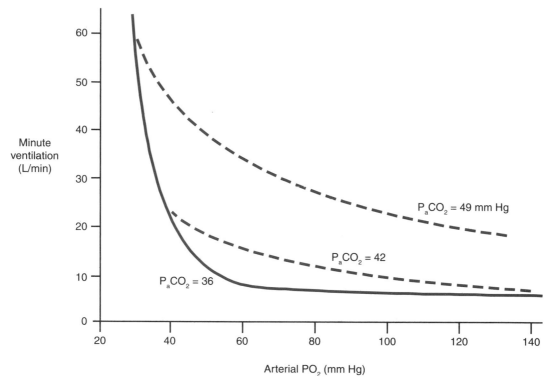

FIGURE 6-6 Ventilatory response to hypoxemia. Minute ventilation is plotted as a function of P_aO_2. Note that ventilation changes little as P_aO_2 decreases from supranormal levels through the normal range. As P_aO_2 decreases below 60 mm Hg, ventilation starts to increase, and the rate of increase becomes marked when P_aO_2 is below 40 mm Hg. The same degree of hypoxemia, when combined with acute hypercapnia, produces even more marked increases in ventilation. (Adapted from Leff A, Schumacker P: *Respiratory Physiology: Basics and Applications,* 1st ed. Philadelphia: WB Saunders, 1993, p. 114.)

be quite close, but in diseased lungs, there may be a range of alveolar PO_2 values, but there can be only one arterial PO_2; the ventilatory response is dependent on the P_aO_2 level and is most accurately represented as a reflection of arterial PO_2).

In keeping with the activity of the peripheral chemoreceptor, as described previously, ventilation changes little as the P_aO_2 decreases from 95 to 60 mm Hg. At that point, the ventilation starts to increase (Fig. 6-6).

THOUGHT QUESTION 6-5: Why and how does an increase in ventilation help someone with acute hypoxemia? What physiological mechanism is responsible for improved oxygenation with an increase in ventilation?

At moderate degrees of hypoxemia—P_aO_2 between 45 and 60 mm Hg, for example—the ventilation is only elevated to approximately twice the normal level. It is only after the P_aO_2 decreases below 40 mm Hg that ventilation increases sharply. When acute hypercapnia is present simultaneously with acute hypoxemia (this occurs typically when a patient has disease of the ventilatory pump or gas exchanger of such severity that the increased stimulation of the controller from the hypoxemia alone is insufficient to lead to hypocapnia), there is a synergistic effect, and ventilation is substantially elevated. The acute hypercapnia produces further stimulation of the peripheral and central chemore-ceptors. This is in contrast to the counterbalancing effect that acute hypocapnia, which normally accompanies the hyperventilation associated with hypoxemia, would have on the ventilatory response to hypoxemia.

> **?** **THOUGHT QUESTION 6-6:** Thinking in terms of evolutionary advantage, why does the ventilatory response to hypoxemia not show a increase in ventilation until the P_aO_2 decreases below 60 mm Hg?

THE VENTILATORY RESPONSE TO HYPERCAPNIA

In contrast to the ventilatory response to hypoxemia, ventilation increases linearly with acute increases in P_aCO_2. The normal range of the response is between 2 and 5 L/min increase in ventilation for each 1-mm Hg increase in P_aCO_2 (Fig. 6-7).

The slope of the ventilatory response appears to be genetically determined. Family members tend to be similar in their location within the normal range, that is, they all tend to have relatively brisk or blunted responses.

The ventilatory response to acute hypercapnia is more pronounced than the response to hypoxemia. Whereas a decline in the P_aO_2 from 90 to 45 mm Hg may cause the ventilation

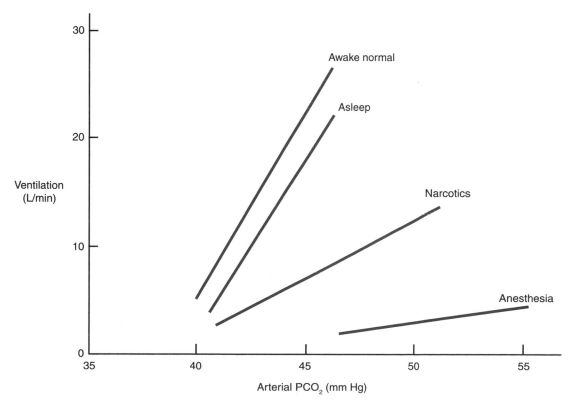

FIGURE 6-7 The ventilatory response to hypercapnia. The ventilatory response to hypercapnia is linear in nature. The normal range of the response is an increase of 2 to 5 L/min in ventilation for each 1 mm Hg increase in P_aCO_2. When a patient is asleep, the curve is shifted to the right. When a patient is under the effects of anesthesia or narcotics, both of which inhibit the response of the controller, the curves are shifted to the right, and the slope is diminished.

to double from 5 to 10 L/min, an increase in the P_aCO_2 from 40 to 50 mm Hg may cause ventilation to increase from 5 to 40 L/min. The reason for this observation is not known, but with our knowledge of respiratory system physiology, we can speculate about a few possibilities. First, with acute hypoxemia, the increase in ventilation leads to hypocapnia and a blunting of the activity of the central chemoreceptor. In contrast, both the peripheral and central chemoreceptors increase their activity with acute hypercapnia. Second, acute hypercapnia results in the accumulation of acid in the blood, which further stimulates the peripheral and central chemoreceptors.

The magnitude of the ventilatory response to hypercapnia can be affected by a number of other factors. When one is asleep, for example, the curve is shifted to the right (less of an effect on ventilation). Similarly, drugs that depress the central nervous system such as narcotics and anesthetics depress the response.

Exercise and Ventilatory Control

We started this chapter by asking you, "Why do you breathe?" You have been learning that the answer to that question is more complicated than you may have thought originally. Now we ask you, "Why do you breathe more when you exercise?" A person with a normal cardiovascular system does not become hypoxemic during exercise, nor does she become acutely hypercapnic. As you might have guessed, the answer to this question is also a bit complicated.

Ventilation increases during exercise in three phases. The first phase is sometimes called the *neurological phase*. Next, one sees a *metabolic phase*. The third phase may be considered the *compensatory phase*.

The initial increase in ventilation occurs almost instantaneously with the onset of exercise. This increase is too quick to be explained by changes in metabolism. It is possible that this first phase is partly a learned response, what some call an anticipatory increase in ventilation. Data from animal studies suggest a possible role for stimulation of the controller by information arising in joint and muscle receptors in the limbs. Passive movement of animals' limbs that simulates exercise has been shown to lead to an increase in ventilation. Because of these hypothesized mechanisms, this first phase of the increase in ventilation has been termed the neurological phase.

As exercise continues, ventilation increases linearly with the increase in oxygen consumption and carbon dioxide production that results from the increased physical activity. Thus, we call this phase of exercise ventilation the metabolic phase. As noted, P_aO_2 and P_aCO_2 remain normal during this phase of exercise, and we refer to the increase in ventilation as *exercise hyperpnea*. Although the increase in ventilation is tightly associated with the changes in oxygen consumption and carbon dioxide production, we do not know how the controller monitors these metabolic changes. For a time, scientists hypothesized the presence of receptors within the pulmonary vasculature that might have the capability of responding to changes in the flux of carbon dioxide returning from the extremities. Definitive proof of these receptors, however, has remained elusive.

As the intensity of exercise continues, the energy needs of the muscles outstrip the ability of the cardiovascular system to supply oxygen for aerobic metabolism. Cells increasingly shift to anaerobic metabolism to supplement their needs. A byproduct of anaerobic metabolism is the production of lactic acid. The acid lowers the pH of arterial blood, which then stimulates the peripheral and central chemoreceptors (with the caveat that the response of the central chemoreceptor may be delayed, given the time it takes for the hydrogen ions to diffuse into the CSF, as discussed previously). Ventilation now increases at a rate faster than the increase in oxygen consumption. The incremental increase in ventilation is a compensation for the development of metabolic acidosis (see Chapter 7) and, consequently, we term this phase of exercise ventilation the compensatory phase. The level of oxygen consumption above which anaerobic metabolism leads to an accumulation of lactic acid in the blood is called the anaerobic threshold and, in normal individuals, it occurs between 40% and 60% of the maximal oxygen consumption (see Chapter 9).

> **?** **THOUGHT QUESTION 6-7:** Children with congenital central hypoventilation syndrome (CCHS) have a functional abnormality of the central chemoreceptor such that it does not respond to elevation in P_aCO_2 or decrease in pH. What would you expect the ventilation to be during exercise in these individuals compared with in normal people?

Ventilatory Control in Acute on Chronic Respiratory Failure

One definition of respiratory failure is the inability to breathe in a manner that supports the basic metabolic needs of the body. If alveolar ventilation is not adequate for the carbon dioxide production of the body, the individual will have an elevated P_aCO_2. If the condition that leads to respiratory failure is chronic and irreversible, then the patient may experience chronic hypercapnia. Acutely, hypercapnia leads to an increase in ventilation as a result of the stimulation of the chemoreceptors, primarily the central chemoreceptor. Within a few days, however, the pH of the blood and the fluid bathing the brain approaches normal levels, and the ventilation comes back down. Because of the increased levels of buffers now present in the blood and brain, further increases in P_aCO_2 have an attenuated effect on ventilation.

Patients with chronic respiratory failure often have problems with hypoxemia as well, with a P_aO_2 that may be near 60 mm Hg. In the clinical setting in which the P_aO_2 decreases acutely (and the P_aCO_2 may increase above the baseline chronic levels) because of a respiratory infection or some other cardiopulmonary process, we describe the situation as *acute on chronic respiratory failure*. The classic teaching about such patients (and you will likely encounter physicians who will tell you this is true) is that because of the chronic hypercapnia and the resulting buffering of the blood and brain, these patients depend on their "hypoxic drive to breathe." The classic teaching also suggests that if you give such a patient supplemental oxygen during an acute problem such as bronchitis or pneumonia, "they will stop breathing" because the stimulation of the peripheral chemoreceptors resulting from hypoxemia will have been relieved. Unfortunately, this teaching, which developed as an explanation for the increase in P_aCO_2 observed with the administration of supplemental oxygen in these patients, is wrong. As you now realize, the controller is more complicated than merely responding to hypoxemia or hypercapnia.

Michelle Aubier, in a very informative study performed in 1980, examined patients with chronic respiratory failure and chronic hypercapnia who presented with an acute respiratory illness and acute hypoxia (M. Aubier et al., *American Review of Respiratory Diseases*, 1980;122:747). Dr. Aubier and his colleagues asked these patients to inhale gas with an FIO_2 of 1.0, or 100% oxygen. On average, the P_aCO_2 increased by approximately 20 mm Hg with the supplemental oxygen, but these patients did *not* stop breathing! Why did the P_aCO_2 increase this much?

In this and subsequent studies, the data showed that the increase in P_aCO_2 with supplemental oxygen could be attributed to three factors. First, there is a small decrease in ventilation as the hypoxemia is relieved. After 15 minutes on the supplemental oxygen, the ventilation was approximately 7% lower than before the administration of oxygen. Second, ventilation/perfusion mismatch is worsened by the administration of oxygen. Recall that poorly ventilated alveoli have a low P_aO_2, which leads to hypoxic pulmonary vasoconstriction of the pulmonary arterioles leading to those alveoli. Blood is sent to better-ventilated alveoli. With the administration of supplemental oxygen, however, oxygen diffuses into the poorly ventilated alveoli, P_AO_2 increases, and the hypoxic pulmonary vasoconstriction is reversed. More blood now goes to what are still poorly ventilated alveoli, and P_aCO_2 increases. This factor accounts for approximately 40% of the acute elevation in P_aCO_2 seen in these patients. The third factor responsible for the increase in P_aCO_2 in these patients is the Haldane effect (see Fig. 5-7). With the administration of oxygen and the increase in P_aO_2, carbon dioxide is displaced from hemoglobin and enters the liquid portion of the blood, resulting in a higher P_aCO_2 (Quick Check 6-1).

QUICK CHECK 6-1	SUPPLEMENTAL OXYGEN AND ACUTE ON CHRONIC RESPIRATORY FAILURE CAUSES OF THE INCREASE IN P_aCO_2

Worsening ventilation/perfusion mismatch
Haldane effect
Reduced ventilation

A 20-mm Hg increase in the P_aCO_2 is not good for the patient because it leads to acute acidosis and, depending on the original P_aCO_2, may bring the level of hypercapnia near the point where it will have an anesthetic effect on the brain. However, acute hypoxemia can be a life-threatening problem and must be treated with supplemental oxygen. Use the lowest amount of oxygen that is necessary to raise the P_aO_2 to approximately 60 mm Hg (oxygen saturation of 90%), and remember that it will not cause the patient to stop breathing.

PUTTING IT TOGETHER

A 45-year-old woman comes to the emergency department complaining of increasing shortness of breath for the past 3 days. She has a history of breast cancer treated with a lumpectomy and radiation. She denies having a cough or fever, and she has no history of heart disease. On physical examination, she appears anxious and in mild respiratory distress with the use of accessory muscles of ventilation. The respiratory rate is 32/min, and her breaths are shallow. The chest examination is notable for a marked decrease in breath sounds over the left hemithorax. An arterial blood gas shows a P_aO_2 = 65 mm Hg, P_aCO_2 = 33 mm Hg, and pH = 7.45. A chest radiograph shows a large pleural effusion (i.e., fluid in the pleural space) filling most of the left chest; there is no shift of the mediastinum, which indicates that the left lung is collapsed under the fluid. A nurse asks you if you want to administer oxygen to slow down the patient's respiratory rate. What do you want to do?

Multiple factors are contributing to the increased respiratory rate and hyperventilation in this patient, but hypoxemia is not one of them. With a P_aO_2 above 60 mm Hg, the patient's oxygen level is not sufficiently low to stimulate the peripheral chemoreceptors to any significant degree. At least two factors are affecting volitional control in this case. The patient is uncomfortable and anxious, both of which typically result in an increase in ventilation. Pulmonary compliance is reduced because the left lung is largely collapsed and virtually all air must go to the right lung. Thus, for any given tidal volume, the alveoli in the right lung are more distended and operate on a flatter portion of the pressure–volume curve than in the absence of the effusion. Rapid, shallow breathing requires less work than slow, deep breathing under these conditions.

The collapse of the left lung leads to stimulation of RARs and decreased stimulation of SARs, both of which activate the respiratory control centers in the medulla. With a large effusion, the chest wall is typically displaced outward, which may lead to stimulation of tendon organs and muscle spindles, resulting in further activation of the central control centers.

Although the administration of oxygen to this patient would not hurt her (and might help reduce her ventilation if the oxygen were administered with a nasal cannula, which would lead to the stimulation of nasal receptors that have an inhibitory effect on the respiratory centers), it will not significantly alter the activity of the controller. The patient needs a thoracentesis, a procedure that drains the fluid from the pleural space with re-expansion of the left lung and restoration of the chest wall to its normal position.

Summary Points

- The control of breathing involves both volitional and automatic elements and is far more complicated than merely reflecting a response to hypoxemia or hypercapnia.
- Hyperventilation describes breathing that is in excess of what is needed to meet the metabolic needs of the body, as reflected in the production of carbon dioxide.
- Hypoventilation describes breathing that is insufficient to meet the metabolic needs of the body, as reflected in the production of carbon dioxide.
- Hyperpnea refers to increased breathing that matches the metabolic needs of the body.
- Tachypnea signifies an increase in respiratory rate above the normal range.
- The neural structures responsible for the automatic control of breathing reside primarily in the medulla. Neurons in the pons may contribute to the transitions from inspiration to expiration.
- Minute-to-minute breathing in a normal person is the consequence of the automatic activity of neurons in the brainstem, which is referred to as the central pattern generator.
- Nerves that innervate the diaphragm exit the spinal cord between the third and fifth cervical vertebrae. The intercostal muscles are supplied by nerve roots emanating from the thoracic spinal cord.
- The breathing of children with congenital central hypoventilation syndrome suggests that in the awake state, breathing is not dependent on a functioning central pattern generator.
- Anxiety and discomfort affect the volitional control of breathing. Patients typically adopt breathing patterns that seem to reduce the discomfort of breathing.
- There are multiple sources of information from receptors throughout the respiratory system that affect the control of breathing.
- Mechanoreceptors in the upper airway, lungs, and chest wall are activated by mechanical distortion of their local environment.
- Stimulation of flow receptors in the airways has an inhibitory effect on ventilation.
- Inflation of the lungs, which stimulates SARs, has an inhibitory effect on the controller. Deflation of the lungs has a stimulatory effect on the controller, mediated via both slowly and rapidly adapting receptors (decreased activity of SARs; increased activity of RARs).
- Stimulation of J receptors (C fibers) by increased pulmonary vascular pressures or interstitial edema may contribute to the tachypnea seen in individuals with CHF.
- Information from Golgi tendon organs and muscle spindles in the chest wall may play an important role in load compensation.
- The peripheral chemoreceptors are located in the carotid bodies and are stimulated by low P_aO_2, high P_aCO_2, and low arterial pH.
- The central chemoreceptor is located in the medulla and is activated by high P_aCO_2 and low arterial pH.

- The differential permeability of the blood-brain barrier for carbon dioxide and hydrogen ions results in a counterbalancing role for the central chemoreceptor relative to the peripheral chemoreceptor when the latter is responding to acute changes in arterial pH.
- The ventilatory response to hypoxia is relatively flat until the P_aO_2 decreases below 60 mm Hg.
- The ventilatory response to hypercapnia is linear.
- The ventilatory response to exercise consists of three phases: neurological, metabolic, and compensatory. None of these depends on the presence of hypoxemia or hypercapnia.
- Patients with chronic hypercapnia do not depend on hypoxemia in order to breathe. Administration of oxygen to patients with acute respiratory problems superimposed on chronic respiratory failure will not cause the patients to stop breathing in most situations. P_aCO_2 increases under these circumstances as a result of a small decrease in ventilation, worsening ventilation/perfusion mismatch, and the Haldane effect.

Answers TO THOUGHT QUESTIONS

6-1. In the absence of activity from the VRG, the muscles of the upper airway would not contract during inspiration. The pressure in the extrathoracic airway is negative during inspiration. In contrast, the pressure surrounding the airway is atmospheric pressure, which is 0 mm Hg by convention. Therefore, a negative transmural pressure (greater outside than inside) occurs across the airway during inspiration. Without contraction of the muscles in the walls of the upper airway to stabilize it during inspiration, the airway narrows, resulting in an increase in airway resistance and, for any given alveolar pressure, a lower flow.

6-2. In an elegant experiment, Dr. John Orem taught cats to hold their breath. By ringing a bell and then spraying a mist of ammonia (a noxious substance to inhale) in the face of a cat, Orem conditioned the cat so that eventually, at the sound of the bell, the cat would hold its breath. Having taught the cat to hold its breath, he then repeated the experiment after placing electrodes into the cat's medulla, thereby allowing him to monitor the activity of the inspiratory neurons. When the cat held its breath, the activity of the inspiratory neurons ceased. These results suggest that we have the ability to volitionally stop the activity of the inspiratory neurons in the central pattern generator (at least for a time; eventually, as P_aCO_2 increases, the inspiratory activity will break through the inhibition, and the diaphragm will begin to contract again). Thus, suppressing the activity of the inspiratory neurons appears to be the mechanism for holding your breath, rather than using the expiratory muscles.

6-3. Flow receptors are present in the upper airway that, when stimulated, provide information to the brain about the response of the respiratory pump to a message coming from the ventilatory controller. Flow receptors in the airways are essentially temperature receptors. In a manner analogous to the calculation of the "wind chill" on a winter day, the body assesses flow from the decrease in temperature during inspiration. Assuming air temperature of 68°F, for example, the inspired gas quickly cools the flow receptors, which are at 98.6°F and 100% humidity. On a hot and humid day in the summer, however, the temperature and humidity of the inspired gas may approximate that of the local environment of the airway receptors. Consequently, it will seem as if inspiratory flow is close to 0. It is possible that this information, which conveys a sense of inadequacy of the response of the ventilatory pump to the output of the controller (termed *neuromechanical* or *efferent–afferent dissociation*, which is discussed in greater detail in Chapter 8) may account for the shortness of breath experienced by some patients under these conditions.

6-4. Because the peripheral chemoreceptors respond to P_aO_2, not oxygen content, anemia alone will not stimulate ventilation at rest. A person who is anemic with a completely normal gas exchanger will have a normal P_aO_2 but reduced oxygen content in the blood. During exercise, ventilation may be increased more than usual in a person with anemia, but the stimulus to the controller that accounts for this is not well delineated. There may be receptors in the peripheral muscles, called *metaboreceptors* or *ergoreceptors*, that monitor the local cellular environment in the muscles. Reduced oxygen content can lead to earlier development of anaerobic metabolism and the accumulation of acid in the tissue, which may stimulate these receptors and lead to the increased ventilation one observes in this setting.

6-5. When a patient hyperventilates, the gas in the alveolus is exchanged more rapidly with the atmosphere. This has the effect of decreasing the alveolar PCO_2 and increasing the alveolar PO_2 compared with the nonhyperventilated condition (recall the alveolar gas equation from Chapter 5). If P_aO_2 is higher, for any given A-aDO_2, the P_aO_2 will be higher.

6-6. The oxygen saturation of the blood is greater than 90% for P_aO_2 values greater than 60 mm Hg. If people responded to mild decreases in P_aO_2 (e.g., from 90 to 80 mm Hg) by increasing ventilation, then for partial pressures above 60 mm Hg, this might be viewed as "wasted ventilation." The person would be doing extra muscle work to increase ventilation without greatly changing hemoglobin saturation and the oxygen content of the blood. When the P_aO_2 decreases below 60 mm Hg, oxygen saturation and oxygen content of the blood decrease sharply. Hyperventilation under these circumstances is an important compensatory mechanism to preserve oxygen delivery to the tissues.

6-7. Despite the absence of a functional central chemoreceptor, children with CCHS have an essentially normal ventilatory response to exercise. The pattern of breathing is a bit more irregular than in normal persons, but the rate of increase in ventilation is remarkably similar. This suggests that the central chemoreceptor is not critical to the controller's response to exercise.

chapter 6

Review Questions

DIRECTIONS: Each of the numbered items or incomplete statements in this section is followed by answers or by completions of the statement. Select the ONE lettered answer or completion that is BEST in each case.

1. A 25-year-old woman comes to your office with a complaint of shortness of breath and a tingling sensation in her fingers and around her mouth. She has no history of heart or lung disease. She appears quite anxious. Her respiratory rate is 8 breaths/min, and she is taking very big breaths. An arterial blood gas on room air shows P_aO_2 = 115 mm Hg, P_aCO_2 = 25 mm Hg, and a pH of 7.52. You would describe her breathing as:

    A. tachypnea
    B. hyperventilation
    C. hyperpnea
    D. hypoventilation

2. You are taking care of a patient with emphysema and chronic airflow obstruction who tells you that his breathing feels better when he uses supplemental oxygen delivered through a nasal cannula (tubing that directs the oxygen into the nose), even though his P_aO_2 without the oxygen is 75 mm Hg and his oxygen saturation is 95%. He does not feel the same effect if you give him the oxygen with a mask. The most likely physiological explanation for this finding is:

    A. oxygen content of the blood is significantly increased
    B. carbon dioxide levels are reduced
    C. receptors in the nose are being stimulated by the flow of oxygen
    D. placebo effect

3. You see a patient in the emergency department who presents with a cough, fever, and shortness of breath that just developed today. You notice that his respiratory frequency and tidal volume appear to be elevated and suspect that his controller is being stimulated. You check an arterial blood gas and find P_aO_2 = 50 mm Hg, P_aCO_2 = 30 mm Hg, and pH = 7.48. Two days later, his P_aO_2 is still 50 mm Hg, and his fever is unchanged. What do you predict the output of the controller to be at this time compared with when you first saw the patient?

    A. increased
    B. decreased
    C. the same
    D. it depends on his diet

4. You are taking care of two patients with hypoxemia. The first patient is a 25-year-old woman with pneumonia. The second patient is a 60-year-old man with emphysema and acute bronchitis. They both have a P_aO_2 of 50 mm Hg. The woman has a P_aCO_2 of 32 mm Hg, and the man has acute hypercapnia with a P_aCO_2 of 46 mm Hg. The male patient's minute ventilation is higher than the young woman's. Which physiological mechanism (or mechanisms) accounts for this difference?

    A. Acute decrease in P_aCO_2 in the female patient causes carbon dioxide in the fluid surrounding the brain to diffuse into the blood, thereby reducing the activity of the central chemoreceptor.

    B. Gender differences in the ventilatory response to hypoxemia are at work.

    C. Acute hypercapnia stimulates the peripheral and central chemoreceptors in the male patient.

    D. Emphysema increases the ventilatory response to hypoxemia.

    E. The Haldane effect is at work.

    F. A and C

    G. all of the above

5. A patient comes into the emergency department with chest pain and shortness of breath for the past 2 hours. He had a myocardial infarction (heart attack) 6 months ago. On physical examination, he is breathing at 28 breaths/min. His electrocardiogram shows evidence of a new infarction. The chest radiograph reveals interstitial edema. The arterial blood gas shows a P_aO_2 of 70 mm Hg, a P_aCO_2 of 34 mm Hg, and a pH of 7.46. The patient's respiratory rate and hyperventilation can be explained by which of the following?

    A. hypoxemia

    B. hypercapnia

    C. stimulation of chemoreceptors

    D. stimulation of pulmonary receptors

    E. discomfort and pain

    F. A, B, and C

    G. D and E

    H. all of the above

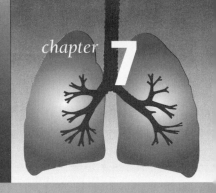

chapter **7**

The Controller and Acid–Base Physiology: An Introduction to a Complex Process

CHAPTER OUTLINE

DEFINITIONS
CARBON DIOXIDE–CARBONIC
 ACID–BICARBONATE BUFFER SYSTEM
THE KIDNEYS AND ACID ELIMINATION
THE PRIMARY ACID–BASE DISORDERS
• Respiratory Acidosis
• Respiratory Alkalosis
• Metabolic Acidosis
• Metabolic Alkalosis

COMPENSATORY MECHANISMS FOR PRIMARY
 ACID–BASE DISORDERS
ANALYZING ACID–BASE DISORDERS: AN
 INITIAL APPROACH
PUTTING IT TOGETHER
SUMMARY POINTS

LEARNING OBJECTIVES

- To describe the role of the respiratory system as a cause of primary acid–base distur-
  bances and as a critical component of the body's ability to compensate for metabolic
  disturbances in acid–base balance.
- To demonstrate the importance of the carbon dioxide–carbonic acid equilibrium in the
  body's acid–base balance.
- To describe the principles underlying the Henderson-Hasselbalch equation and their
  application to acid–base physiology.
- To enumerate the four basic acid–base disorders and the use of the arterial blood gas
  (ABG) test in recognizing the disturbances.
- To introduce a few simple elements of renal physiology that are important for the under-
  standing of acid–base physiology.

The normal function of enzymatic activity of the body requires that the concentration of
hydrogen ion in the blood, and consequently the cells, be closely regulated. Hydrogen ion
concentration in the blood is very low, approximately 40 neq/L, and these ions are small
and highly reactive, that is, they bind more strongly to negatively charged molecules than
do other cations such as sodium and potassium. The range of hydrogen ion concentration
in the blood that is compatible with human life is 16 to 126 neq/L, which translates into a
blood pH between 7.80 and 6.90. Major problems occur, however, in multiple physiologic
processes when pH deviates below 7.20 or above 7.55. Proteins necessary for many of the
chemical reactions in the body are sensitive to relatively small changes in blood pH. The
binding of oxygen to hemoglobin, as you recall from Chapter 5, for example, is diminished
in an acid environment.

Carbon dioxide dissolved in blood combines with water to form carbonic acid, which is in equilibrium with hydrogen ions and bicarbonate.

$$CO_2 + H_2O \Leftrightarrow H_2CO_3 \Leftrightarrow H^+ + HCO_3^-$$

Regulation of P_aCO_2 is, therefore, critical to the maintenance of an acceptable pH level in the blood. From this relationship, you can also see that derangements of the respiratory system, including any of the three components—controller, pump, and gas exchanger—we have discussed thus far that lead to an elevated or reduced P_aCO_2, can have significant implications for the health of the organism. In Chapter 6, we discussed how changes in pH and P_aCO_2 affect the peripheral and central chemoreceptors. The controller is designed, in part, to respond to alterations in blood pH and to prevent significant changes in hydrogen ion concentration associated with elevations in carbon dioxide production.

A full understanding of acid–base physiology requires integrated study of the respiratory and renal systems. Because all important journeys must begin with one step, we will start your investigation of this complex topic by examining the important role of the respiratory system in acid–base physiology.

Definitions

An **acid** is a substance that can donate a hydrogen ion (proton). A **base** is a substance that can accept a hydrogen ion. The production of carbon dioxide resulting from the metabolism of carbohydrates leads, as already noted, to the formation of carbonic acid, which dissociates into a hydrogen ion and molecule of bicarbonate.

$$CO_2 + H_2O \Leftrightarrow H_2CO_3 \Leftrightarrow H^+ + HCO_3^-$$

Normal metabolism, primarily of carbohydrates and fats, leads to the production of approximately 10,000 to 15,000 mmol of carbon dioxide every day. We sometimes refer to carbonic acid as a **volatile acid** because we can eliminate it by breathing more, essentially driving the reaction above to the left by increasing alveolar ventilation, thereby decreasing PCO_2. In contrast, the body makes a relatively small amount of **fixed acids**, mostly phosphates and sulfates resulting from the metabolism of protein, which must be eliminated via the kidneys. Every day, approximately 50 to 100 meq of fixed acids are produced as a consequence of the body's metabolism. Absent the respiratory system and its ability to eliminate carbon dioxide, the body would quickly be overwhelmed by acid. Absent functioning kidneys, the body would also be overwhelmed by acid, but it would take days rather than minutes for that to occur.

A **buffer** is a molecule that is able to accept or release hydrogen ions so that changes in the free hydrogen ion concentration and, hence, the pH, are minimized. The primary intracellular buffers are proteins, phosphates, and hemoglobin in red blood cells (RBCs). Serum albumin and bone also serve as important sites for the buffering of acid. Sodium and potassium ions on the surface of the bone may exchange with protons, or bone mineral may be dissolved, a process that leads to the release of buffers into the extracellular fluid. The primary source of buffer in the extracellular fluid is bicarbonate, largely because of the ability of the body to eliminate carbon dioxide and drive the carbonic acid equilibrium to the left.

A process that leads to a disturbance in the balance of acids or bases in the body is noted with the suffix *-osis*. Thus, a process leading to the excess production of acid is an **acidosis**. A process leading to excess production of base is an **alkalosis**. If the process that alters the relative concentration of acids and bases alters the pH of the blood, the suffix *-emia* is used to signify the effect on arterial hydrogen ion concentration. If the pH of the blood is higher than normal (i.e., more basic, with a pH > 7.44), an **alkalemia** is present; if the pH is lower than normal (increased concentration of hydrogen ions, with a pH < 7.36), an **acidemia** exists. It is important to remember that the presence of a process leading to a change in the quantities of acid and base in the system does not necessarily imply a change in the pH of the blood. The buffering capacity of the body and its ability to initiate secondary processes to counter or compensate for the initial challenge to the acid–base equilibrium may result in a normal pH despite the primary physiological disorder.

Carbon Dioxide–Carbonic Acid–Bicarbonate Buffer System

As noted, the primary extracellular buffer is bicarbonate via its role in the carbon dioxide–carbonic acid–bicarbonate system. In its gaseous form, carbon dioxide dissolves in the aqueous portion of blood to form carbonic acid. The amount of carbon dioxide that dissolves in the blood is proportional to the partial pressure of carbon dioxide in solution, that is, the P_aCO_2, which in a normal person is the same as the alveolar PCO_2. At normal body temperature (37°C), the amount of carbon dioxide that dissolves in blood is:

$$[CO_2]_{dissolved} = 0.03 PCO_2$$

Assuming a normal arterial PCO_2:

$$[CO_2]_{dissolved} = 0.03(P_aCO_2) = 0.03(40) = 1.2 \text{ mmol/L}$$

Some of the dissolved carbon dioxide combines with water to form carbonic acid. Given the chemical equilibrium between dissolved carbon dioxide and carbonic acid:

$$CO_2 + H_2O \Leftrightarrow H_2CO_3$$

You can use the dissolved carbon dioxide as a marker for the amount of carbonic acid in the blood.

The *Henderson-Hasselbalch equation* defines the dissociation relationship for weak acids:

$$pH = pKa + \log \frac{[base]}{[acid]}$$

For the carbon dioxide–carbonic acid system, this can be written:

$$pH = pKa + \log \frac{[HCO_3^-]}{0.03(PCO_2)}$$

Note that in this equation, the relationship for dissolved CO_2 is used where the concentration of acid is found. This convention is used because the concentration of carbonic acid is difficult to measure, and PCO_2, which is easily measured, is in equilibrium with carbonic acid. The ratio of dissolved CO_2 to carbonic acid at body temperature is approximately 400 to 1. The pKa of carbonic acid is 3.5. When adjusted for the use of dissolved CO_2 as a marker for carbonic acid, the relationship becomes:

$$pH = 6.1 + \log \frac{[HCO_3^-]}{0.03(PCO_2)}$$

To achieve a normal pH in blood, the ratio of $[HCO_3^-]$ to the amount of dissolved carbon dioxide in the blood must equal 20 (log of 20 = 1.3). A normal bicarbonate concentration is 24 meq/L, and a normal P_aCO_2 is 40 mm Hg, which results in a ratio of 20.

In the end, we have a gas, carbon dioxide, which, while not an acid itself, leads to the production of carbonic acid once dissolved in water. Problems that lead to an increase in P_aCO_2 result in the accumulation of acid. Conversely, as acid builds up in the system and is buffered by bicarbonate to form carbonic acid, the respiratory system offers an incredibly fast and efficient system for eliminating the acid. By hyperventilating and lowering P_aCO_2, the body causes the equilibrium to shift to the left, allowing further buffering to occur. This response is made possible because of the physiology of the controller, as discussed in Chapter 6. The increase in hydrogen ion concentration stimulates the chemoreceptors, which trigger an increase in ventilation. Whereas nonbicarbonate buffers are limited by the quantity of the buffer and the pH, the capacity of the carbon dioxide–carbonic acid–bicarbonate system is determined primarily by the concentration of bicarbonate. The ability to modify pH by changing ventilation thus enhances the effectiveness of this buffering system.

The Kidneys and Acid Elimination

It is beyond the scope of this book to detail the physiology of the kidneys and their role in acid–base homeostasis. However, a full understanding of the workings of the body requires one to integrate the physiology of multiple organ systems (you will see this again in Chapter 9, in which we introduce concepts of cardiovascular physiology in our exploration of exercise). Therefore, we will highlight a few of the major physiologic processes by which the kidneys contribute to acid–base balance.

The process of normal living requires the generation of energy and the metabolism of a mixture of carbohydrates, fats, and proteins. As already discussed, consumption of carbohydrates and fats leads primarily to the production of carbon dioxide, which is in equilibrium with carbonic acid. In a typical adult diet, we also metabolize protein, which results in the production of an additional 50 to 100 meq of noncarbonic acid every day. These hydrogen ions must be excreted in the urine.

The kidneys are designed to remove chemicals from the blood via filtration in the glomerulus and secretion in the tubule and to reabsorb some of these filtered molecules and ions at other portions of the tubule to maintain chemical, water, and acid–base balance. Bicarbonate is filtered in the glomerulus and must be reabsorbed to avoid lowering the serum bicarbonate level. Approximately 90% of bicarbonate reabsorption occurs in the proximal tubule, and the remainder occurs more distally in the nephron. To remove the acid load resulting from the metabolism of protein, the kidneys must secrete hydrogen ion in the tubule. After the hydrogen ions are secreted into the lumen, they must bind with urinary buffers, primarily phosphate (HPO_4^{2-}) and ammonia (NH_3). Thus, most metabolic acid is eliminated as $H_2PO_4^-$ and NH_4^+; there is relatively little free hydrogen ion eliminated via the urine. Disorders of the glomerulus or the tubule can disrupt these processes and impair the ability of the kidneys to eliminate acid.

THOUGHT QUESTION 7-1: You are experimenting with two new drugs. One of the drugs will block the ability of the respiratory system to eliminate carbon dioxide. The other drug will prevent the kidney from eliminating hydrogen ions in the urine. If you give each drug to different animals and measure the pH one hour later, which drug will result in a more severe acidosis?

The Primary Acid–Base Disorders

There are four primary acid–base disorders that disturb normal physiology (Table 7-1). They may occur alone (simple disorders) or in combination (mixed disorders). As you will see in the next section, these acid–base processes may also occur in order to compensate for a primary abnormality. In these cases, the initial process is known as the primary disturbance, and the secondary process is known as the compensation. Two of the primary disorders (respiratory acidosis and alkalosis) are the consequence of problems with the respiratory system, and two (metabolic acidosis and alkalosis) are the consequence of problems with oxygen delivery to the tissues, abnormalities in metabolic processes, ingestion of toxins, and diseases of the kidneys. A full discussion of these disorders is beyond the scope of this book, but we will touch on the basic concepts.

RESPIRATORY ACIDOSIS

Respiratory acidosis is characterized by an increased P_aCO_2, a decreased pH, and ultimately, a mild increase in the serum bicarbonate concentration (remember: an increase in P_aCO_2 drives the carbonic acid reaction to the right, which results in an increase in serum bicarbonate, even before any compensation). A decrease in alveolar ventilation leads to an increase in alveolar PCO_2 and, subsequently, an increase in P_aCO_2. As we have just discussed, the increase in dissolved carbon dioxide ultimately leads to the formation of carbonic acid and a decrease in pH (Fig. 7-1).

 Use Animated Figure 7-1 to play a primary respiratory acidosis and observe how the change in P_aCO_2 affects the carbonic acid equilibrium and, subsequently, the pH. The picture of the "balance" or "scale" is used to illustrate the changing equilibrium between carbonic acid and bicarbonate. In essence, this is a visual representation of the modified Henderson-Hasselbalch equation presented earlier in the chapter.

$$pH = 6.1 + \log \frac{[HCO_3^-]}{0.03(PCO_2)}$$

Note that PCO_2 is used in this portion of the diagram as a marker for carbonic acid, as discussed previously.

The hydrogen ion produced by virtue of a respiratory acidosis cannot be buffered by bicarbonate, the primary extracellular buffer.

$$H_2CO_3 + HCO_3^- \Leftrightarrow HCO_3^- + H_2CO_3$$

As the reaction shows, we would merely take a proton from the molecule of carbonic acid (producing a molecule of bicarbonate) and give it to the bicarbonate (producing a molecule of carbonic acid); there is no change in the concentration of either molecule, and no change in pH results from this exchange. Instead of relying on extracellular bicarbonate,

TABLE 7-1 Primary Acid–Base Disorders			
TYPE OF DISORDER	**CHANGE IN P_aCO_2**	**CHANGE IN SERUM BICARBONATE**	**CHANGE IN pH**
Respiratory acidosis	↑		↓
Respiratory alkalosis	↓		↑
Metabolic acidosis		↓	↓
Metabolic alkalosis		↑	↑

Note: The changes shown reflect acute alterations in P_aCO_2 and serum bicarbonate level without invoking buffering or compensatory mechanisms.

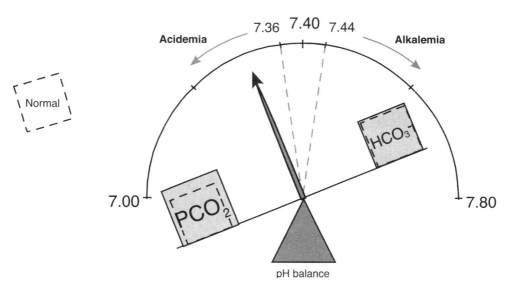

Primary respiratory acidosis
(shown before compensation)

$$\uparrow PCO_2 = K \frac{\dot{V}CO_2}{\downarrow \dot{V}_A} \quad \text{where} \quad \begin{array}{l} \dot{V}CO_2 = \text{carbon dioxide production} \\ \\ \dot{V}_A = \text{alveolar ventilation} \end{array}$$

$$CO_2 + H_2O \rightleftharpoons \underset{\text{Carbonic acid}}{H_2CO_3} \rightleftharpoons H^+ + \underset{\text{Bicarbonate}}{HCO_3^-}$$

FIGURE 7-1 Respiratory acidosis. When the P_aCO_2 increases because of hypoventilation, there is a shift to the right in the carbonic acid equilibrium (shown) and pH decreases (along with a small increase in serum bicarbonate level). Note that PCO_2 is used as a marker for carbonic acid on the visual "balance" shown here. This is because the PCO_2 is easier to measure and is in equilibrium with carbonic acid. We are assuming a constant level of carbon dioxide production (approximately 200 mL/min). **Also see Animated Figure 7-1.**

the protons produced by respiratory acidosis must be buffered by proteins, particularly the hemoglobin in RBCs.

$$H_2CO_3 + Buf \Leftrightarrow HBuf + HCO_3^-$$

The bicarbonate produced by this reaction may then diffuse into the serum, resulting in an increase in the serum bicarbonate.

An increase in P_aCO_2 (i.e., hypoventilation) can result from problems with the controller, ventilatory pump, or gas exchanger. For example, a person who attempts to commit suicide by taking an overdose of a sedative medication that depresses the ventilatory controller will develop respiratory acidosis. Mild to moderate derangements in the ventilatory pump or gas exchanger, such as seen in individuals with mild asthma, typically do not cause a respiratory acidosis by themselves because the controller responds to the accumulation of carbon dioxide and hydrogen ions by increasing ventilation and restoring a normal or near-normal P_aCO_2. On the other hand, depression of the controller is associated with respiratory acidosis, even in the presence of a normal ventilatory pump and gas exchanger.

In the setting of a normal controller, severe abnormalities of the gas exchanger or ventilatory pump may lead to hypercapnia and respiratory acidosis because maximal achievable ventilation under these conditions may not be adequate to achieve alveolar ventilation sufficient to meet the carbon dioxide production associated with the metabolic state of the person. A person with severe emphysema and acute pneumonia, for example, may develop hypercapnia despite the best efforts of the controller to increase ventilation.

RESPIRATORY ALKALOSIS

Respiratory alkalosis is characterized by a decreased P_aCO_2, an increased pH, and a mild decrease in the serum bicarbonate concentration. An increase in alveolar ventilation leads to a decrease in alveolar PCO_2 and subsequently a decrease in P_aCO_2. The decreased P_aCO_2 drives the carbonic acid–carbon dioxide equilibrium to the left (see equation on page 150); the hydrogen ion concentration decreases and pH increases (Fig. 7-2). Use Animated

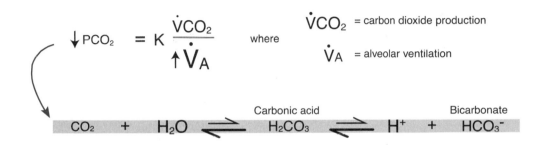

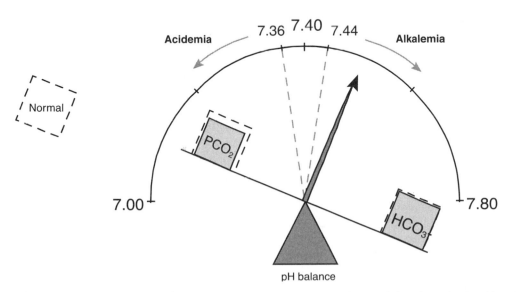

FIGURE 7-2 Respiratory alkalosis. When the P_aCO_2 decreases, there is a shift to the left in the carbonic acid equilibrium (shown) and pH increases (along with a small decrease in serum bicarbonate level). As in Figure 7-1, note that PCO_2 is used as a marker for carbonic acid on the visual "balance" shown here. This is because the PCO_2 is easier to measure and is in equilibrium with carbonic acid. We are assuming a constant level of carbon dioxide production (approximately 200 mL/min). **Also see Animated Figure 7-2.**

Figure 7-2 to play a primary respiratory alkalosis and observe how the change in P_aCO_2 affects the carbonic acid equilibrium and subsequently the pH.

Because one would expect the decrease in P_aCO_2 and the increase in pH to reduce the activity of the respiratory centers in the brainstem, the presence of a primary respiratory alkalosis suggests that other sources of information or stimuli are affecting the controller. Patients with mild, acute asthma attacks, for example, may be short of breath and anxious and experience stimulation of pulmonary receptors as a consequence of the bronchospasm and inflammation of the airways. All of these factors can contribute to an increase in ventilation and produce the respiratory alkalosis that is typical of this condition.

THOUGHT QUESTION 7-2: A 16-year-old, previously healthy young woman is brought to you complaining of a sensation of not being able to get a deep breath. She is light-headed and has tingling in her fingers. One hour ago, she learned that her father was killed in an automobile accident. Her alveolar–arterial oxygen gradient is normal, as is her lung examination and chest radiograph. Her P_aCO_2 is 24 mm Hg, and her pH is 7.53. How would you characterize her acid–base disturbance, and what do you think is causing it?

METABOLIC ACIDOSIS

Metabolic acidosis is characterized by a reduced bicarbonate concentration and a low pH. It is generally accompanied by compensatory hyperventilation. Metabolic acidosis is typically the consequence of the accumulation of fixed acids in the body (e.g., caused by renal failure, accumulation of lactic acid from anaerobic metabolism, or ingestion of toxins) or the loss of bicarbonate from the kidneys or gastrointestinal (GI) tract (e.g., as is seen with diseases of the renal tubule or profuse diarrhea). As fixed acids accumulate, they are initially buffered by bicarbonate, lowering the bicarbonate concentration (Fig. 7-3). Use Animated Figure 7-3 to play a primary metabolic acidosis and observe how the change in bicarbonate affects the carbonic acid equilibrium and subsequently the pH (shown before respiratory compensation). The presence of an increased hydrogen ion concentration stimulates the chemoreceptors, increasing ventilation and leading to a decrease in P_aCO_2 and further buffering of the acid (a process that is one of the compensatory mechanisms discussed later in this chapter).

The physiologic derangements that lead to metabolic acidosis can be subdivided into two major categories: those that lead to an elevated **anion gap** (referred to as anion gap acidoses) and those that are not associated with an elevated gap (referred to as non–anion gap acidoses). The anion gap is the difference in concentrations between the commonly measured *anions* and *cations* in the blood. It can be altered by certain acids, so calculation of this difference serves as an aid to recognition and diagnosis of metabolic acidosis. The chemicals in the blood must maintain electrical neutrality, and the cations (positively charged ions) such as sodium, potassium, calcium, and magnesium are balanced by anions (negatively charged molecules) such as chloride, bicarbonate, proteins, sulfates, and phosphates.

The anion gap is the difference between the concentration of the major cation, sodium, and the anions that we routinely measure, chloride and bicarbonate (Fig. 7-4). A normal anion gap is between 5 and 11 meq/L (note: some people calculate the anion

Primary metabolic acidosis
(shown without compensation)

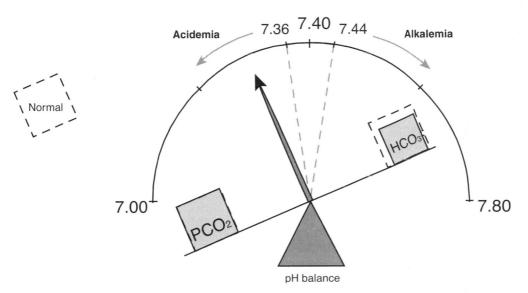

FIGURE 7-3 Metabolic acidosis. When fixed acids accumulate in the body or when bicarbonate is lost, there is a reduction in bicarbonate concentration, resulting in a shift to the right in the carbonic acid equilibrium (shown) and a decrease in pH. The situation is shown before respiratory compensation (discussed in the text; as depicted, carbon dioxide resulting from the buffering of the acid is being eliminated as fast as it is produced). As in Figures 7-1 and 7-2, note that PCO_2 is used as a marker for carbonic acid on the visual "balance" shown here. **Also see Animated Figure 7-3.**

gap by including potassium as one of the cations; in this case, the normal values for the gap are between 9 and 15 meq/L). A significant portion of the anion gap is composed of the negatively charged proteins such as albumin. Therefore, when we speak of a "normal" range for the anion gap, we are assuming a normal level of albumin. If the serum albumin is low, as is seen in malnourished individuals and in people who have disease in which albumin is lost from the urine, the normal range for the anion gap must be adjusted downward (2.5 meq/L for every 1 g/dL decline in the serum albumin concentration).

A limited number of conditions produce an anion gap acidosis, including renal failure (in which reduced filtration capability of the kidney leads to the accumulation of sulfates and phosphates from the metabolism of proteins), hypoperfusion of tissues leading to lactic acid accumulation, uncontrolled diabetes leading to ketoacidosis, and the ingestion of drugs such as aspirin and toxins such as ethylene glycol. Although the respiratory system can compensate, to a degree, for the change in hydrogen ion concentration that results from these processes, elimination of the unmeasured anions requires a functioning kidney.

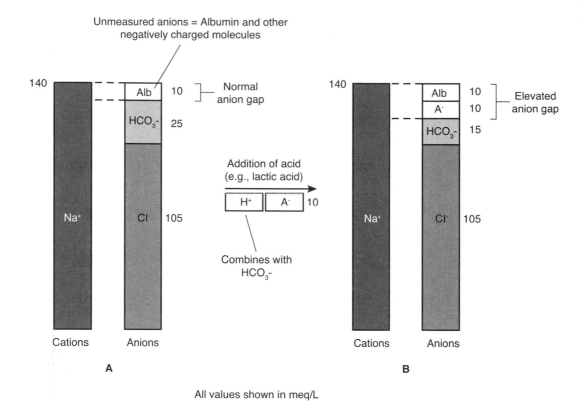

All values shown in meq/L

FIGURE 7-4 Anion gap. **A,** The anion gap is formed by the difference between the measured cations (Na^+) in the blood and the measured anions (Cl^- and HCO_3^-). Negatively charged proteins such as albumin account for this difference under normal circumstances. **B,** Addition of fixed acids may reduce the serum bicarbonate level and lead to an elevation of the anion gap.

Non–anion gap acidosis is most commonly caused by conditions associated with a loss of bicarbonate from the body. The classical clinical example of this is moderate to severe diarrhea (the fluid lost from the GI tract has a high concentration of bicarbonate). Disorders of the kidney, such as renal tubular acidosis, are also characterized by non–anion gap acidosis.

METABOLIC ALKALOSIS

Metabolic alkalosis is characterized by an increased bicarbonate concentration and an elevated pH (Fig. 7-5). Use Animated Figure 7-5 to play a primary metabolic alkalosis and observe how the elevation in bicarbonate concentration affects the carbonic acid equilibrium and subsequently the pH (shown before respiratory compensation). This disorder is generally accompanied by compensatory hypoventilation. Metabolic alkalosis is often the consequence of loss of hydrogen ion from the GI system (e.g., as is seen with prolonged vomiting). It is also seen in association with diuretic therapy, with contraction of the total amount of fluid in the body and loss of potassium from the blood, as well as with the intake of excessive bicarbonate (as seen with the use of antacids). For metabolic alkalosis in general, as the pH increases, stimulation of the chemoreceptors is reduced, and hypoventilation generally follows.

Primary metabolic alkalosis
(shown without compensation)

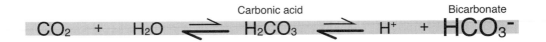

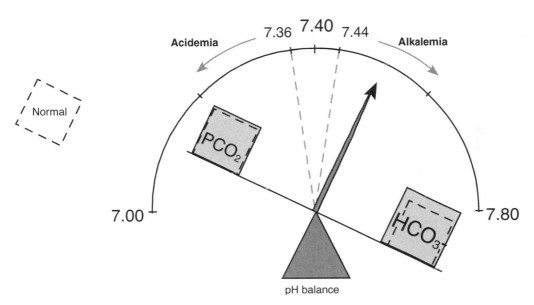

FIGURE 7-5 Metabolic alkalosis. When the serum bicarbonate concentration increases (e.g., because of vomiting or diuretic abuse), it results in a shift to the left in the carbonic acid equilibrium (shown) and an elevation in pH. The situation is shown before respiratory compensation (discussed in the text). As in other figures in this chapter, note that PCO_2 is used as a marker for carbonic acid on the visual "balance" shown here. **Also see Animated** Figure 7-5.

Compensatory Mechanisms for Primary Acid–Base Disorders

The body must maintain the pH of the blood and cells within a narrow range. To accomplish this, the respiratory and renal systems provide compensatory mechanisms for the primary acid–base disturbances we have just discussed. If a problem occurs in either of these systems, the other system can adjust its function to offset the problem. In the simplest version of this process, a primary respiratory acidosis, for example, is offset by a secondary (compensatory) metabolic alkalosis. Conversely, a metabolic acidosis is offset by a secondary respiratory alkalosis (Table 7-2).

With a primary respiratory acidosis, the P_aCO_2 increases and the pH decreases. As already noted, the protons in this disorder are initially buffered by intracellular proteins. This metabolic compensation results in the production of bicarbonate that diffuses back into the serum, resulting in an increase in the serum bicarbonate. If the respiratory acidosis persists, the kidneys begin to compensate by increasing the amount of hydrogen ions eliminated in the urine, a process that leads to the generation of bicarbonate that enters

TABLE 7-2 Primary Acid–Base Disorders and Compensatory Mechanism

PRIMARY DISORDER	PRIMARY DISTURBANCE	COMPENSATORY MECHANISM	COMPENSATORY CHANGE
Respiratory acidosis	P_aCO_2 ↑	Metabolic alkalosis	Serum bicarbonate ↑
Respiratory alkalosis	P_aCO_2 ↓	Metabolic acidosis	Serum bicarbonate ↓
Metabolic acidosis	Serum bicarbonate ↓	Respiratory alkalosis	P_aCO_2 ↓
Metabolic alkalosis	Serum bicarbonate ↑	Respiratory acidosis	P_aCO_2 ↑

the blood. In the case of a primary respiratory alkalosis, the P_aCO_2 decreases, the pH increases, and the kidneys compensate by excreting more bicarbonate. In the example of a primary metabolic acidosis, a fixed acid accumulates, which causes the pH to decrease. The respiratory system compensates by increasing ventilation, and P_aCO_2 decreases. Finally, with a primary metabolic alkalosis, bicarbonate concentration increases, pH increases, and the respiratory system compensates by reducing ventilation, which results in an increase in P_aCO_2 (Fig. 7-6).

   Use Animated Figure 7-6 to select a primary acid–base disturbance and observe the effect on the pH caused by the change in PCO_2 or bicarbonate. Then choose the secondary (compensatory) process and note how it serves to move the pH back toward normal.

The goal of these compensatory mechanisms is to restore the pH of the blood to normal or near-normal levels. However, the compensation never results in a pH that is all the way back to 7.40, with the exception of the metabolic compensation for a chronic respiratory alkalosis. If the pH is fully restored to 7.40 or beyond, you should look for the presence of two, simultaneous, primary acid–base disturbances (more on how to recognize this in the next section).

Respiratory compensation for metabolic acid–base disorders can occur within seconds to minutes. The sensitivity of the chemoreceptors and the timely response of the controller to changes in their output ensure this. Metabolic compensation for respiratory disorders, however, generally requires 2 to 5 days to be fully evident. The kidneys' ability to adjust levels of hydrogen ion and bicarbonate is not as rapid as the respiratory response.

Analyzing Acid–Base Disorders: An Initial Approach

In the initial approach to simple acid–base disturbances, you must decide if the patient has one or more primary problems and whether the disorder (or disorders) are acute or chronic, that is, whether compensation has occurred (Table 7-3). The data upon which we rely to make these determinations are the ABG results, which tell us the pH and the P_aCO_2, and the serum electrolyte levels, which provide us with the serum bicarbonate and allow us to calculate the anion gap.

The first step is to examine the pH. If the pH is outside the normal range (7.36–7.44), an acidemia or alkalemia is present. Next you must determine what process, or -osis, is accounting for the abnormal pH. Look at the P_aCO_2—is it in the direction that you would expect if there were a primary respiratory process to account for the change in pH? For example, if the pH is 7.30, you know that the patient has an acidemia. When you look at the rest of the ABG results, you note that the P_aCO_2 is 53 mm Hg (significantly above a normal value of 40 mm Hg), indicative of respiratory acidosis. Thus, the change in the

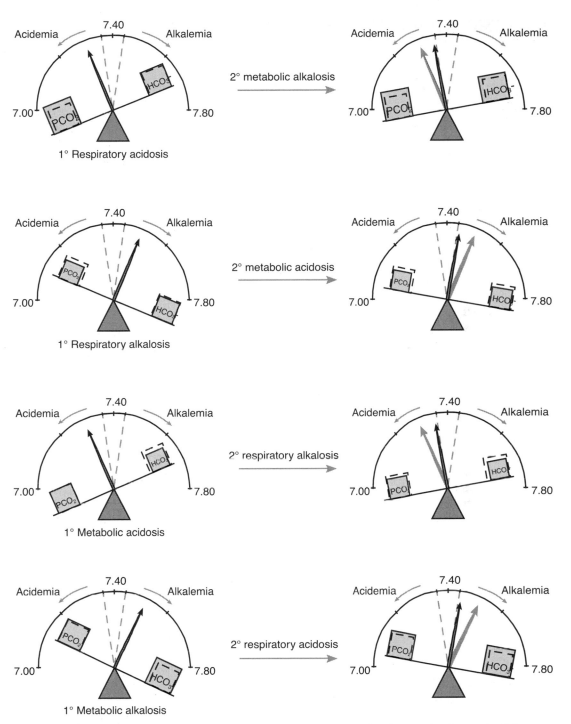

FIGURE 7-6 Primary disorders and compensations. The balance analogy is used to illustrate the changes in PCO$_2$, serum bicarbonate, and pH that occur with each of the primary acid–base disturbances and the expected compensation by either the renal or respiratory system. **Also see Animated Figure 7-6.**

TABLE 7-3 Approach to Acid–Base Disturbances: Key Questions

1. Is the pH abnormal? If so, in what direction?
2. Is the change in P_aCO_2 in the direction expected for a primary respiratory disturbance? (If the P_aCO_2 changes in the opposite direction from the pH, a primary respiratory disorder is present.)
3. If a primary respiratory disturbance is present, is it acute or chronic?
4. If a primary metabolic disturbance is present, is an abnormal anion gap present?
5. If a primary metabolic disturbance is present, is there an appropriate respiratory system response?

P_aCO_2 in this case is in the direction that indicates that the primary disturbance is consistent with a respiratory acidosis contributing to an acidemia. (Note: when there is a primary respiratory system cause of an acid–base abnormality, the P_aCO_2 moves in the opposite direction of the pH; in this example of a primary respiratory acidosis, the pH is down from normal, and the P_aCO_2 is elevated.)

Now you must make a calculation to determine if the respiratory acidosis is acute or chronic. The pH change associated with an acute respiratory acidosis is greater than for a chronic respiratory acidosis because the kidney has not yet had time to compensate for lowered pH by retaining more bicarbonate and eliminating protons (as ammonium NH_4^+). With acute respiratory acidosis, the pH decreases approximately 0.08 units for every increase in P_aCO_2 of 10 mm Hg. In contrast, a chronic respiratory acidosis is associated with a decrease in the pH of approximately 0.03 units for every 10 mm Hg increase in the P_aCO_2. The serum bicarbonate also changes acutely as a reflection of initial buffering of the change in pH, and to a greater degree, chronically, as the kidneys compensate for the acute respiratory disorder (see Table 7-3).

If there is evidence of an abnormal pH but the P_aCO_2 has moved in a direction that is opposite of what you expect for a primary respiratory disorder, then you are dealing with a primary metabolic disturbance with respiratory compensation. For example, if the pH is 7.30, an acidemia is present. If the P_aCO_2 is 28 mm Hg, the acidemia cannot be attributable to a primary respiratory process; the patient is hyperventilating (which, with no other disturbance present, would tend to make the blood more alkalemic). This means that the patient has a primary metabolic acidosis with a respiratory compensation. You would then look at the patient's serum electrolytes and calculate the anion gap to determine if the metabolic acidosis is in the anion gap or non–anion gap category. With primary metabolic disturbances, the respiratory system, if it is normal, can compensate almost instantaneously. Thus, having identified that there is a primary metabolic abnormality, the question you must address is whether there is an appropriate respiratory system response rather than whether the problem is acute or chronic (Table 7-4). For metabolic acidosis, the appropriate respiratory system response results in a decrease in P_aCO_2 of 1.2 mm Hg for every 1 meq/L decrease in the concentration of bicarbonate in the blood. In patients with metabolic alkalosis, the P_aCO_2 should increase 0.7 mm Hg for every 1 meq/L increase in bicarbonate concentration.

Remember that abnormalities in any of the components of the respiratory system—the controller, ventilatory pump, or gas exchanger—may prevent an appropriate response to a primary metabolic disturbance.

We have presented an approach that is useful in understanding simple acid–base disturbances. You must remember, however, that it is possible for two or three primary abnormalities to be present simultaneously. In these circumstances, your clue to a complex acid–base disturbance is that the simple rules we have described do not seem to

TABLE 7-4 Primary Acid–Base Disturbances and Associated Compensations		
PRIMARY DISTURBANCE	**PRIMARY CHANGE**	**COMPENSATORY RESPONSE**
RESPIRATORY ALKALOSIS	$\downarrow P_aCO_2$	
Acute		2 meq/L reduction in serum bicarbonate concentration for every 10 mm Hg decrease in P_aCO_2; pH increases 0.08 units for every 10 mm Hg decrease in P_aCO_2
Chronic		4 meq/L reduction in serum bicarbonate concentration for every 10 mm Hg decrease in P_aCO_2; pH increases 0.03 units for every 10 mm Hg decrease in P_aCO_2
RESPIRATORY ACIDOSIS	$\uparrow P_aCO_2$	
Acute		1 meq/L increase in serum bicarbonate concentration for every 10 mm Hg increase in P_aCO_2; pH decreases 0.08 units for every 10 mm increase in P_aCO_2
Chronic		4 meq/L increase in serum bicarbonate concentration for every 10 mm Hg increase in P_aCO_2; pH decreases 0.03 units for every 10 mm Hg increase in P_aCO_2.
METABOLIC ALKALOSIS	$\uparrow HCO_3$	
		0.7 mm Hg increase in P_aCO_2 for every 1 meq/L increase in serum bicarbonate concentration
METABOLIC ACIDOSIS	$\downarrow HCO_3^-$	
		1.3 mm Hg decrease in P_aCO_2 for every 1 meq/L decrease in serum bicarbonate concentration

explain the data before you. Your understanding of the physiology of the respiratory, renal, and GI systems will then guide you to making correct diagnoses of your patient's problem.

THOUGHT QUESTION 7-3: A 25-year-old woman with a history of asthma comes to the emergency department complaining of shortness of breath for the past 3 hours. She has had stomach cramps and vomiting for 3 days and has been unable to eat or drink very much. Today she visited her friend's house, where a cat was present. The patient is allergic to cats, and she developed acute bronchospasm that has been resistant to her bronchodilator inhalers. An ABG shows a high pH (7.58), a low P_aCO_2 (33 mm Hg), and a high serum bicarbonate (30 meq/L). How would you describe her acid–base status?

PUTTING IT TOGETHER

You are evaluating a 55-year-old man who comes to you with shortness of breath. The patient has a history of moderate emphysema. He has had diarrhea for the past month after going camping in New Hampshire. The patient admits to drinking water out of several streams, and a recent evaluation of his stool shows the presence of a parasite. As his diarrhea has worsened, his breathing has become more difficult. He does not have a cough or a fever. On physical examination, the patient is breathing at 22 breaths/min with mild use of accessory muscles of ventilation. The chest reveals mildly reduced breath sounds with an inspiratory to expiratory ratio of 1 to 1.5 (when a healthy individual takes deep breaths during examination of his lungs, the ratio of inspiration to expiration is 1 to 1; a prolonged expiratory phase of the respiratory cycle usually indicates the presence of airway obstruction). There are no wheezes. The abdominal examination shows mild diffuse tenderness.

The patient's chest radiograph shows evidence of increased lung volume (total lung capacity [TLC]) but no pneumonia. Pulmonary function tests are unchanged from his last visit to you, at which time he had no significant respiratory complaints. You obtain an ABG, which shows a P_aO_2 of 75 mm Hg, P_aCO_2 of 28 mm Hg, and pH of 7.32. The serum bicarbonate is reduced at 14 meq/L (normal, 23–27 meq/L). The anion gap is 8. What do you make of the acid–base disturbances in this individual? What is the relationship between his diarrhea and his shortness of breath?

You determine that the patient has a primary metabolic acidosis from the loss of bicarbonate in his diarrhea. The pH is below the normal range, which indicates the presence of an acidemia. The P_aCO_2, however, is low, which suggests that the acidemia is attributable to a metabolic process, and the respiratory system is attempting to compensate for the metabolic disturbance. The respiratory system has responded normally despite his emphysema (increased dead space), but the abnormalities of the gas exchanger and ventilatory pump (hyperinflation) result in a significant increase in the work of breathing in order to achieve the necessary degree of hyperventilation. Therefore, in this case, treatment of the patient's respiratory symptoms requires eradication of his underlying parasitic infection and a reduction in the diarrhea or administration of supplemental bicarbonate to make up for the losses in the diarrhea.

Summary Points

- The body maintains close control of the blood pH to ensure optimal function of enzymatic reactions within the cells.
- The relationship between dissolved carbon dioxide and carbonic acid and the consequent ability of the respiratory system to affect the acidity of the blood are critical to maintenance of an appropriate blood pH.
- Although the respiratory system is able to compensate for increased production of carbonic acid, elimination of fixed acids produced by the metabolism of proteins requires a functioning renal system.
- The primary intracellular buffers, which accept or release hydrogen ions to minimize changes in pH in the setting of an acidosis or alkalosis, are proteins, phosphates, and hemoglobin in RBCs. The primary source of buffer in the extracellular fluid is bicarbonate.
- The terms *acidosis* and *alkalosis* refer to processes that lead to a reduction or increase in serum pH, respectively.
- The terms *acidemia* and *alkalemia* refer to the status of the blood pH—below normal in the case of acidemia and above normal in the case of alkalemia.

- The four primary acid–base disorders are respiratory acidosis, respiratory alkalosis, metabolic acidosis, and metabolic alkalosis.
- Respiratory acidosis is characterized by an increased P_aCO_2, a decreased pH, and a mild increase in the serum bicarbonate concentration.
- Respiratory alkalosis is characterized by a decreased P_aCO_2, an increased pH, and a mild decrease in the serum bicarbonate concentration.
- Metabolic acidosis is characterized by a reduced bicarbonate concentration and a low pH. Conditions that lead to a metabolic acidosis may be divided into two categories: those associated with a normal anion gap and those that result in an increased anion gap.
- Metabolic alkalosis is characterized by an increased bicarbonate concentration and an elevated pH.
- Disorders of the respiratory and renal systems, when they lead to primary acid–base disturbances, generate compensatory processes in the other system to minimize the effect of the primary problem on the blood pH. Whereas respiratory compensation can occur within seconds to minutes, renal compensation may take 2 to 5 days before it is complete.
- The analysis of an acid–base disturbance requires a systematic approach that includes an assessment of the blood pH, the P_aCO_2, serum bicarbonate level, and the anion gap.
- Because of the delay in the response of the renal system to primary acid–base abnormalities of the respiratory system, one can categorize respiratory system acid–base disorders as acute or chronic. In an acute respiratory system disturbance, the pH changes by 0.08 units for every change of 10 mm Hg in the P_aCO_2; for a chronic disorder, the pH changes by 0.03 units for every change of 10 mm Hg in the P_aCO_2.
- If there is evidence of an abnormal pH but the P_aCO_2 has moved in a direction that is opposite of what you expect for a primary respiratory disorder, then you are dealing with a primary metabolic disturbance with respiratory compensation.
- A complete understanding of acid–base disorders requires integration of respiratory, renal, and GI system physiology.

Answers TO THOUGHT QUESTIONS

7-1 The animal receiving the drug that blocks carbon dioxide elimination will have a much more severe acidosis and, in fact, will be dead. Remember that the lungs eliminate nearly 15,000 mmol of carbon dioxide every day, which, if not eliminated, quickly transforms into carbonic acid. In contrast, the kidneys eliminate only 50 to 100 meq of acid every day. One hour after blocking renal excretion of acid, there will be a minor effect on hydrogen ion concentration. Both organ systems are necessary for maintenance of life, but the role of the respiratory system in acid–base balance is more critical for minute-to-minute adjustments in pH.

7-2 The patient has a primary respiratory alkalosis. The P_aCO_2 is low, which indicates that she is hyperventilating. The pH is higher than normal, consistent with an alkalemia caused, in this case, by the fact that the low P_aCO_2 is driving the carbonic acid–carbon dioxide equilibrium to the left. The patient has no apparent problem with the gas exchanger or ventilatory pump. It is most likely that her respiratory alkalosis is the result of hyperventilation brought on by an acute emotional stress. Remember that behavioral factors can also stimulate the controller via the cerebral cortex.

7-3 In this case, there is more than one primary disorder present. You first look at the pH, which is high. This tells you that an alkalemia is present. You next look at the P_aCO_2. It is low, consistent with a respiratory alkalosis (which is, in this case, a consequence of the acute bronchospasm and stimulation of the controller by virtue of the patient's breathing discomfort and possibly stimulation of stretch receptors in the lung). However, the pH is too high for an acute respiratory alkalosis, and the serum bicarbonate concentration is elevated, when you would have expected it to decrease in the setting of a respiratory alkalosis. These findings can be explained by the presence of a primary metabolic alkalosis that is the result of vomiting (loss of hydrogen ion from the stomach) and volume depletion from the inability to eat or drink (a reduction in the body's fluids contributes to an increase in the concentration of the serum bicarbonate).

Review Questions

DIRECTIONS: Each of the numbered items or incomplete statements in this section is followed by answers or by completions of the statement. Select the ONE lettered answer or completion that is BEST in each case.

1. A 25-year-old man in a coma is brought into the emergency department by ambulance. He fell from a construction site and has a head injury. The respiratory rate is 6 breaths/min, and the chest examination and chest radiograph findings are normal. You predict that the patient's acid–base status will be:

    A. respiratory acidosis
    B. respiratory alkalosis
    C. metabolic acidosis
    D. metabolic alkalosis

2. You are taking care of two patients with severe renal failure who are waiting for dialysis. One of the patients is 24 years old and has no respiratory problems. The other is 62 years old and has severe emphysema. Which patient is likely to have a lower pH?

    A. the 24-year-old patient
    B. the 62-year-old patient

For questions 3 to 6, determine the acid–base abnormality. Assume the patient had the following normal values before the disorder began:

P_aCO_2 = 40 mm Hg
Serum bicarbonate = 24 meq/L
pH = 7.40

3. ABG: P_aCO_2 = 75 mm Hg, P_aCO_2 = 50 mm Hg, pH = 7.32; serum bicarbonate = 25 meq/L

    A. acute respiratory acidosis
    B. chronic respiratory acidosis
    C. respiratory alkalosis
    D. metabolic alkalosis

4. ABG: P_aO_2 = 82 mm Hg, P_aCO_2 = 34 mm Hg, pH = 7.37; serum bicarbonate = 19 meq/L

    A. acute respiratory alkalosis
    B   chronic respiratory alkalosis
    C. respiratory acidosis
    D. metabolic acidosis

5. ABG: P_aO_2 = 65 mm Hg, P_aCO_2 = 55 mm Hg, pH = 7.35; serum bicarbonate = 29 meq/L

    A. acute respiratory acidosis
    B. chronic respiratory acidosis
    C. metabolic alkalosis
    D. metabolic acidosis

6. ABG: P_aO_2 = 110, P_aCO_2 = 30 mm Hg, pH = 7.44; serum bicarbonate = 20 meq/L

    A. acute respiratory alkalosis
    B. chronic respiratory alkalosis
    C. metabolic alkalosis
    D. chronic respiratory acidosis

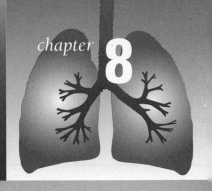

chapter **8**

The Physiology of Respiratory Sensations

CHAPTER OUTLINE

LEARNING OBJECTIVES

- To describe the multiple physiologic sources of respiratory sensations.
- To illustrate links between receptors that are important in ventilatory control and respiratory sensation.
- To describe the sources of sensory information arising from the lung that are important to the perception of breathing.
- To describe the importance of the relationship between the neural output of the controller and the mechanical response of the ventilatory pump in modulating the intensity of respiratory discomfort.
- To demonstrate that dyspnea, or respiratory discomfort, is composed of multiple, qualitatively distinct sensations.
- Using asthma as a model, to demonstrate that within a patient suffering from a single pathologic problem, there may be several physiological sources of breathing discomfort.

Breathe in. Breathe out. We do it every day, all day long. For the most part, we are unaware of our breathing, and if we asked you to describe what it feels like to breathe, you would have to think about the question for a few moments before answering. Certainly, when you exercise hard, you notice your breathing and, if you push yourself to your physiological limits (more on this in Chapter 9), you will likely experience breathing discomfort, or **dyspnea.** If you are among the 5% to 10% of the population with asthma, you undoubtedly have noted abnormal breathing sensations at some point. Are these sensations the same as when you exercise? If not, what are the differing physiological mechanisms responsible for the disparate sensations? The answer to the first question is "no" for the majority of individuals with asthma who describe a sensation of "chest tightness" with their asthma flare-ups. This sensation contrasts with the sensation of "huffing and puffing" or "heavy breathing" most commonly seen with mild to moderate exercise in the absence of respiratory system disease. Studies of respiratory sensations, which are described in this chapter, provide data on the quality of dyspnea in a variety of disease states and physiological conditions and suggest hypotheses regarding the physiological origins of the sensations.

Before you started your study of respiratory physiology, you might have responded to the question, "Why do we get short of breath?" by answering, "Because oxygen levels decrease or carbon dioxide levels increase." You learned in Chapter 6, however, that the control of ventilation is more complex than a simple response to hypoxemia or hypercapnia. Similarly, respiratory sensations and dyspnea cannot be explained fully by changes in P_aO_2 or P_aCO_2, either. Rather, these sensations and dyspnea reflect the processing of neurological information arising from receptors throughout the respiratory system, along with an apparent comparison of the outgoing motor commands from the controller with the mechanical response of the ventilatory pump.

Acknowledging that we are walking a fine line between physiology and pathophysiology in the discussion of respiratory sensations and dyspnea, we believe that this is an appropriate topic for you to consider at this time because it reinforces many of the concepts you have been learning about the controller and the ventilatory pump. In addition, the topic is rarely addressed in pathophysiology texts, and many clinicians do not have a firm understanding of the physiological mechanisms that underlie dyspnea. Finally, an appreciation of the links between physiology and respiratory sensations will make you a better diagnostician when confronted with patients who complain of shortness of breath.

We will approach the physiology of respiratory sensations by using the following framework. First, we will examine the relationships between ventilatory control and respiratory sensations. Second, we will explore the role of the lungs as sensory organs. Finally, we will address the issue of dissociation between outgoing motor command from the controller and the subsequent response of the ventilatory pump. Much of what we have learned about respiratory sensations in recent years comes from studies of the *language of dyspnea*. These studies have developed and used dyspnea questionnaires to systematically elicit the words and phrases that patients use to describe their breathing discomfort[1,2]. In contrast to pain, which includes sensations that we all experience from time to time as part of our lives, even if we are essentially healthy, most healthy people only experience dyspnea with exercise. Thus, we do not grow up with a readily available vocabulary to describe our breathing discomfort when we develop a cardiopulmonary disease that interferes with normal physiology.

THOUGHT QUESTION 8-1: A man brings his 70-year-old mother to see you because she has been "short of breath." He says that her breathing is often "noisy," and she looks like she is "laboring" to breathe when she walks up the stairs. When you question the patient, she says that her breathing is "fine," and she denies having any breathing discomfort. Does she have dyspnea?

Relative to the other elements of respiratory physiology described in this book, the study of the physiology of breathing sensations is fairly new, the concepts we present here are less well substantiated by experimental data and are, therefore, more controversial, and the physicians with whom you work in the future may be unaware of the principles we outline. For these reasons, we will provide several key references in case you wish to explore this topic in more detail in the future or have the opportunity to share your knowledge in a discussion with your colleagues.

Respiratory Sensations and Ventilatory Control

Acute hypoxemia and acute hypercapnia stimulate the chemoreceptors and lead to an increase in ventilation. These changes in P_aO_2 and P_aCO_2 also lead to a sensation most commonly described as "air hunger, urge to breathe, or need to breathe"[3–5]. Because the intensity of the sensations seemed to increase in concert with the change in ventilation, physiologists initially thought the sensation was the consequence of the physical activity of the breathing. More recent studies, however, have demonstrated that these alterations in blood gases can cause air hunger in spinal cord–injured patients maintained on mechanical ventilators[4] as well as in experimental subjects whose ventilation is fixed because of the administration of a paralytic agent[5]. In addition, if you ask a person to voluntarily constrain her breathing while you increase her P_aCO_2 level by administering a mixture of inhaled gases that contains elevated levels of carbon dioxide, the intensity of the discomfort is heightened (In this example, the subject is asked to maintain a constant level of ventilation despite the increased output from the controller associated with acute hypercapnia; in this sense, the ventilation is constrained[6].) The sensation of air hunger, therefore, appears to arise directly from information projected from the chemoreceptors to the sensory cortex because the activity of the ventilatory muscles is unchanged but the intensity of the discomfort is increasing. We will see shortly, however, that the amount of breathing the individual is doing for any given level of hypoxemia or hypercapnia can affect the intensity of the sensation.

As discussed in Chapter 6, other factors may also stimulate the controller. Acute acidemia activates the chemoreceptors and produces a sensation of air hunger even in the absence of changes in blood gases. Pathologic conditions such as pulmonary embolism (blood clots to the lungs), asthma, and heart failure (with increases in pulmonary vascular pressures and leakage of fluid from pulmonary capillaries) also cause an increase in ventilation, probably from stimulation of pulmonary receptors. Although these conditions may be associated with multiple respiratory sensations, patients with these problems may describe a sensation of air hunger, even in the absence of alterations in P_aO_2 or P_aCO_2.

The increase in ventilation associated with exercise is attributable to a number of factors that stimulate the controller in different ways (see Chapters 6 and 9). The breathing sensation associated with exercise, however, is somewhat different than the air hunger noted above. In studies of the respiratory sensations associated with exercise, a sensation of heavy breathing or breathing more appears to be most characteristic[7]. We believe this sensation is an indicator of cardiovascular fitness (more on this in Chapter 9).

Elevated ventilation implies that ventilatory muscle activity has increased. In many cases, this heightened activity is perceived as a sensation of increased work or effort of breathing. The sense of effort is thought to result from a neural message that is sent from the motor cortex to the sensory cortex at the same time as the signal is sent to the muscles to contract. This "copy" of the motor message has been given the name **corollary discharge** and is thought to be a mechanism by which the sensory cortex is able to monitor the activity of the motor cortex (Fig. 8-1).

In the example of a person with a normal ventilatory pump, we are not usually aware of a sense of effort to breathe, much as it does not require significant effort to raise your arm or to walk down the street, activities that require the activation of a number of skeletal muscles. However, if you are asked to lift a 50-lb weight or to walk down the street with legs weakened by polio, the output of the motor cortex and the sense of effort will increase substantially. In most clinical sensations, the work or effort of breathing becomes prominent when there is a problem with the ventilatory pump, such as increased airflow obstruction, a stiff chest wall, or a weakened diaphragm.

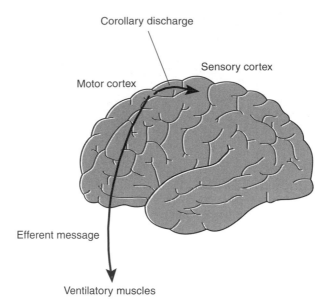

FIGURE 8-1 Corollary discharge and the sense of effort. As efferent (outgoing) neural messages are sent from the motor cortex to the ventilatory muscles, a simultaneous message is sent to the sensory cortex. This second neural signal is called a *corollary discharge* and is thought to be responsible for the sense of effort.

The Lungs As Sensory Organs

Chapter 6 describes the role of a variety of pulmonary receptors in the determination of minute ventilation, tidal volume, and respiratory rate. These receptors also play an important part in the production and relief of breathing discomfort.

Pulmonary stretch receptors (PSRs) appear to allow us to monitor the size of a breath. As discussed previously, the sensation of air hunger produced experimentally by acute hypercapnia is made worse when a subject is constrained to breathe with small tidal volume and is relieved by larger breaths. Presumably, the information on breath size arises from stimulation of the PSRs.

Rapidly adapting receptors (RARs) and C fibers, which are stimulated by chemicals and possibly inflammatory processes in the lungs, also have roles in ventilatory control and conceivably contribute to breathing discomfort associated with airway inflammation. Perhaps the most intriguing example of the possible role of PSRs and RARs in the formation of respiratory sensations is in patients with asthma. Individuals who experience acute bronchoconstriction in the context of asthma may develop shortness of breath even before their lung function tests demonstrate a significant abnormality in expiratory flow. At these mild degrees of bronchoconstriction, the most common sensation described is chest tightness or constriction[8]. Evidence that this sensation arises from the lungs is provided by an experiment in which the dyspnea of asthma is reduced by the inhalation of lidocaine[9], a local anesthetic that can reduce the activity of pulmonary receptors, especially C fibers , and a study in which the sensation has been shown to persist in subjects with bronchoconstriction in whom the work of breathing has been removed by the institution of mechanical ventilation[10].

Stimulation of RARs and C fibers in the lungs may also contribute to the sensation of burning of which many patients with acute bronchitis complain. Anecdotal reports of

patients with pulmonary embolism, or blood clots to the lungs, who received throm-
bolytic therapy to dissolve the clots and experience near-instantaneous relief of their
breathing discomfort as the clots dissolve, suggest that stimulation of pulmonary vascular
receptors by the increased pulmonary artery pressure associated with the embolism may
be the source of the breathing discomfort in this condition.

Efferent–Reafferent Dissociation: The Relationship Between Neural Output from the Controller and the Response of the Ventilatory Pump

When a neural signal is generated from the controller and sent to the ventilatory muscles,
we describe the message as an **efferent neural impulse**. When the ventilatory muscles
contract in response to the signal, a flow of air is generated, the lungs and chest wall expand, and receptors throughout the ventilatory pump are stimulated. Flow, volume dis-
placement, muscle tension, and joint position are all monitored by the body by virtue of
stimulation of receptors in the lungs and chest wall and the messages sent back to the sensory cortex, messages we term **afferent neural impulses**.

When the efferent impulses and the response of the ventilatory pump, as evidenced by
the **reafferent** activity of receptors throughout the lungs and chest wall (termed *reafferent*
because it is in response to the original outgoing motor command), appear to be appropriate, we generally experience little breathing discomfort. However, when there is mismatch
of the outgoing and incoming signals, or **efferent–reafferent dissociation**, the intensity of
sensations such as air hunger and the effort or work of breathing increase. This notion
was first introduced as the concept of length–tension inappropriateness to signify that
under conditions of increased airflow obstruction, the ventilatory muscles generate
greater tension with less shortening than in the normal state[11]. Others have used the term
neuromechanical dissociation to signify the mismatch between the efferent signals from the
controller and the subsequent performance of the ventilatory pump[12]. We prefer the concept of efferent–reafferent dissociation because it is more inclusive[13]. The model allows
for input from a vast array of receptors in the upper airways, lower airways, pulmonary
parenchyma, and chest wall in the modification of the intensity of dyspnea, and there are
increasing experimental data to support this model[14,15].

> **?**
>
> **THOUGHT QUESTION 8-2:** A patient with paralysis from the neck down since a diving
> accident 3 years ago (when he severed his spinal cord at the cervical level, C2) is sus-
> tained chronically on a mechanical ventilator. He develops pneumonia and complains
> of being short of breath. His P_aO_2 is a bit low, and you increase the FIO_2 (the fraction of
> oxygen in the inspired gas) to bring it back to normal. The patient continues to complain
> about breathing discomfort. His P_aCO_2 is normal. The ventilator settings, with the ex-
> ception of the recently increased FIO_2, are the same as usual. Which change in the
> ventilator—an increased tidal volume or an increased respiratory frequency—is more
> likely to alleviate the breathing discomfort?

In our discussion of the controller in Chapter 6, we noted that one of the unique features of the respiratory system is that it was under both voluntary and reflex control.
Without thinking about our breathing, the brainstem respiratory centers ensure that we

breathe in a manner that provides an adequate P_aO_2 and P_aCO_2 and acid–base status. On the other hand, we can volitionally increase or decrease our respiratory rate and tidal volume. To this point, we have discussed the sense of effort (mediated by corollary discharge) in the context of efferent neural output from the motor cortex, which implies a voluntary activity. Is there a similar sense of effort when the reflex components of the controller are stimulated, that is, when the efferent signals are originating in the medulla? An experiment to answer this question used a model in which subjects targeted their breathing to an elevated level of ventilation while their P_aCO_2 levels were altered[16]. This study is important because it is one of the first studies to show that the intensity of two types of respiratory discomfort may vary in opposite directions under the same experimental conditions. The study also provides additional information about the origins of the sense of effort. This finding helps build the case that qualitatively different respiratory sensations arise from different physiological mechanisms.

In this study[16], subjects breathed at a constant level of ventilation while P_aCO_2 was kept at 40 mm Hg or surreptitiously raised to 50 mm Hg (the subjects and the researcher supervising the ratings were blinded to the level of carbon dioxide) by adding different amounts of CO_2 to the inspired gas. Subjects rated two sensations, "effort to breathe" and "breathing discomfort," at two levels of ventilation, resting breathing and at an elevated level determined by a visual target. The goals of the study design were twofold. First, the investigators wished to look at sensations associated with two levels of motor output—normal ventilation and an elevated level of ventilation—under the condition that P_aCO_2 was 40 mm Hg. At a normal P_aCO_2, the neural output to the ventilatory muscles was presumably arising primarily from the motor cortex. One would predict that the sense of effort, which correlates with the neural output from the motor cortex, would increase when the subjects increased their level of ventilation. The results of the study bore out this prediction.

The second part of the protocol was designed to examine sensations associated with an elevated level of ventilation when P_aCO_2 was varied between 40 and 50 mm Hg. The investigators hypothesized that neural output to the ventilatory muscles from the brainstem that results from a *reflex* command, via stimulation of the chemoreceptors, does not generate the same sense of effort as output from the motor cortex associated with a *voluntary* command. If the hypothesis were true, you would predict that the sense of effort associated with a given level of ventilation would be less with an elevated P_aCO_2 than when one has to voluntarily work to generate the elevated level of ventilation. The results demonstrated that, at a P_aCO_2 of 50 mm Hg, the sense of effort was less than when the P_aCO_2 was 40 mm Hg. Thus, despite the fact that the output of the ventilatory pump was the same (i.e., ventilation was constant) under both conditions, the sense of effort was different. This is consistent with the hypothesis that ventilation that is stimulated by the chemoreceptors and originates in the brainstem does not generate the same corollary discharge to the sensory cortex as does voluntary hyperpnea. On the other hand, the sense of breathing discomfort (what other experiments have shown to be "air hunger" or "an increased urge to breathe") increased as P_aCO_2 was increased, despite the fact that ventilation was held constant. Taken together, these findings suggest the dyspnea associated with acute hypercapnia, and presumed stimulation of the chemoreceptors, is physiologically distinct from the sense of effort.

The Dyspnea of Asthma: A Model of Multiple Sensations

To this point, we have discussed sensations that arise from or are modified by stimulation of the controller, from receptors in the lungs, and from dissociation between efferent and reafferent neural information. Table 8-1 summarizes some of the most common qualities

TABLE 8-1 Physiological Sources of Common Respiratory Sensations		
QUALITY OF SENSATION	PHYSIOLOGICAL SOURCE	COMMON CLINICAL EXAMPLES
Air hunger, urge to breathe, need to breathe	Stimulation of controller—chemoreceptors, pulmonary receptors	Acute hypoxemia, hypercapnia, asthma, pulmonary embolism
Chest tightness, constriction	Stimulation of pulmonary receptors	Asthma
Effort or work to breathe	Increased efferent output from the motor cortex	Emphysema, asthma, obesity, diaphragm paralysis
Burning	Stimulation of the pulmonary receptors	Acute bronchitis
Heavy breathing, huffing and puffing	Metaboreceptors or ergoreceptors in the skeletal muscles	Exercise

of breathing discomfort, the disease states or conditions associated with them, and the hypothesized physiological origins of the sensations.

Many cardiopulmonary conditions are associated with several physiological derangements that can lead to multiple sensations and qualitatively distinct types of dyspnea. Attention to the ways that patients describe their respiratory discomfort can provide clues to the underlying physiological problems. Asthma provides an excellent example of the way in which the nature of the physiologic disturbance changes with the severity of the disease process and the associated respiratory symptoms change with the physiology[8].

Asthma is a disease characterized by inflammation of airways. The inflammatory process leads to swelling or edema of the lining of the airways, increased production of mucus, infiltration of the walls of the airways by white blood cells, and the release of chemicals that cause contraction of smooth muscle surrounding the airways. The inflammatory process and the spasm of the airway muscle, termed *bronchospasm*, lead to increased airway resistance. Asthma is a type of obstructive lung disease.

In the mildest forms of asthma, or at the very beginning of an asthma attack, you may see bronchoconstriction with minimal, if any, change in airway resistance or lung function as measured by the forced vital capacity (FVC) and the forced expiratory volume in 1 second (FEV_1). At this point, patients complain of a sensation of chest tightness or constriction, presumably caused by stimulation of pulmonary receptors (possibly RARs and C fibers) from the bronchospasm. Clinicians who are unaware of the significance of chest tightness may think that these individuals are malingering (i.e., making up their symptoms) because the "objective" measures of lung function appear to be normal.

As the asthma progresses in severity, the inflammatory process becomes more evident, bronchospasm worsens, and airway resistance increases. The FEV_1 decreases. There is now a more evident problem with the ventilatory pump. The ventilatory muscles must work harder to generate sufficient negative intrathoracic pressure to overcome the resistance and produce flow. Efferent neural output from the controller increases. The person with the asthma attack now perceives a sensation of increased effort or work of breathing. As the FEV_1 decreases further, the sense of effort increases, although chest tightness continues to be present as well (Fig. 8-2).

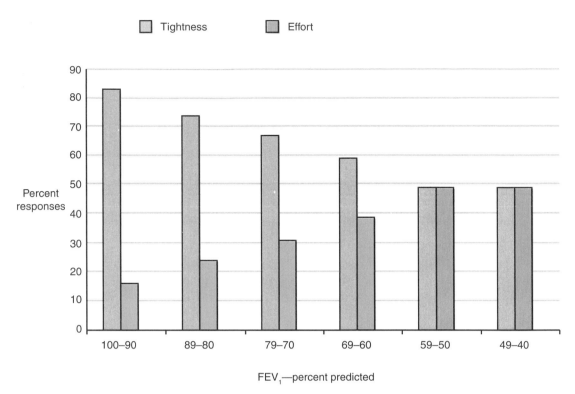

FIGURE 8-2 Evolution of respiratory sensations in asthma. The graph depicts the percentage of responses given by subjects with mild asthma (normal lung function at baseline) in whom acute bronchoconstriction was induced. Note that at the very early stages of bronchospasm, when lung function is still normal or only mildly reduced (a forced expiratory volume in 1 second [FEV_1] greater than 80% of predicted for age and height is considered within the normal range), the sensation of chest tightness predominates. As more severe degrees of airway obstruction develop, the sensation of increased effort or work of breathing is rated as often as chest tightness. (Adapted from Moy ML, Weiss JW, Sparrow D, et al: Quality of dyspnea in bronchoconstriction differs from external loads. *Am J Respir Crit Care Med* 2000, 162:454.)

 THOUGHT QUESTION 8-3: Do patients suffering from an acute asthma attack complain of more difficulty breathing in or out? Why?

With more severe asthma, as the FEV_1 decreases even farther, the patient may begin to report a sensation of air hunger, a type of dyspnea that appears to be associated with stimulation of the controller. Hyperventilation, evidence of stimulation of the controller, is typically seen in mild to moderate asthma. In these stages of asthma, increased levels of ventilation may result from afferent feedback from pulmonary receptors (RARs and C fibers) and from behavioral factors (discomfort and anxiety). In moderate to severe asthma, hypoxemia may also develop, typically from ventilation/perfusion mismatch, which further stimulates the respiratory controller.

In one patient, with the progress of a single respiratory problem, we see physiologic changes in the controller, pump, and gas exchanger, each of which contributes to the development of respiratory sensations that are perceived and reported as symptoms of the disease. With knowledge of the relationship between sensations and physiology (Fig. 8-3), the clinician can gain great insights from the patient's story.

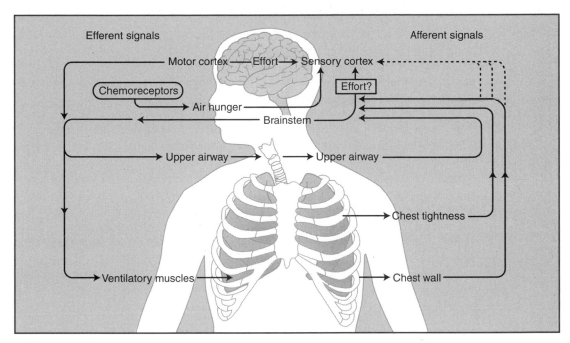

FIGURE 8-3 The physiological pathways of respiratory sensations. The figure summarizes the efferent and afferent pathways that are instrumental in the generation of respiratory sensations. The corollary discharge (see Fig. 8-1) is indicated by the path between the motor and sensory cortex. Output from the brainstem may also contribute to the sense of effort. Manning et al.[14] state: "The sense of air hunger is believed to arise, in part, from increased respiratory activity within the brain stem, and the sensation of chest tightness probably results from stimulation of pulmonary receptors, which transmit information to the brain via the vagus nerve. Although afferent information from airway, lung, and chest-wall receptors most likely passes through the brain stem before reaching the sensory cortex, the dashed lines indicate uncertainty about whether some afferents bypass the brain stem and project directly to the sensory cortex." (Adapted from Manning HL, Schwartzstein RM: Pathophysiology of dyspnea. *N Engl J Med* 1995, 333:1548).

PUTTING IT TOGETHER

You are asked to evaluate a 29-year-old woman for asthma that is not responding to bronchodilator medications. She is an investment banker who has worked 80-hour weeks since graduating from college. With little time for exercise, she has gained 20 pounds since graduation. About 2 months ago, she decided to start jogging and experienced shortness of breath after only going one half mile. She went to her local doctor who concluded that she might have asthma and started her on bronchodilator inhalers to prevent bronchospasm from developing with exercise. She has been taking medication for more than 1 month with no improvement despite the fact that the doses of the drugs have been increased. You are now seeing her for further management of her asthma.

When you speak to the patient, you ask her to describe the breathing discomfort she has with exercise. She states that it feels like heavy breathing or huffing and puffing. The air goes in and out easily. She has never had a sensation of chest tightness or constriction. When she stops to rest, she recovers quickly, usually within 2 or 3 minutes. In fact, her discomfort feels like what she remembers about exercise in high school and college, but it now occurs with much less strenuous activity. You listen to her breath sounds; there are no wheezes. Her lung function is normal. What do you think is causing her breathing discomfort? Does she just need more medication for her asthma?

Based on the patient's description of her breathing discomfort, you might be skeptical that she actually has asthma. She has had no chest tightness or sense of increased effort to breathe. The sensations are similar to what she has had in the past with exercise, they just occur earlier than she expects. You stop all of her medication and three days later perform an exercise test to reproduce the symptoms and to study her physiology. The test shows that she has no bronchospasm with exercise, no hypoxemia, and does not reach a ventilatory limit at the point that she stops because of shortness of breath (more on the physiology of exercise limits in Chapter 9). She does reach her maximal predicted heart rate, consistent with a diagnosis of deconditioning. The patient has not engaged in any consistent aerobic exercise in 7 years and is out of shape.

The source of the breathing discomfort associated with exercise has not been clearly delineated. Since P_aO_2 and P_aCO_2 do not change until very intense levels of exercise, we do not believe that the chemoreceptors are responsible. Some investigators have hypothesized that receptors in the skeletal muscles, at times called metaboreceptors *or* ergoreceptors, *may respond to local changes in the metabolic milieu of the muscle cell and transmit information to the brain that is perceived as breathing discomfort.*

You reassure the patient that her sensations represent normal physiology, not a disease. With a defined exercise program, she is doing much better (taking no medications) when you see her again 2 months later.

Summary Points

- *Dyspnea* is the medical term used to denote breathing discomfort.
- Stimulation of the ventilatory controller leads to a sensation of air hunger, the urge to breathe, or the need to breathe.
- Acute hypoxemia and hypercapnia can cause air hunger, even in the absence of activity of the ventilatory muscles; thus, these sensations do not arise from the muscles or act of breathing. Constraining ventilation can worsen air hunger for a given level of hypoxemia or hypercapnia.
- The dyspnea associated with exercise is commonly described as heavy breathing or breathing more. The receptors responsible for this sensation have not been clearly defined and may reside within the skeletal muscles.
- The sensation of increased work or effort of breathing is probably a reflection of a neural discharge from the motor cortex to the sensory cortex. This corollary discharge provides a "copy" to the sensory cortex of the motor output going to the ventilatory muscles.
- The sense of increased work or effort of breathing is most commonly seen when there is a disorder of the ventilatory pump.
- Stimulation of the pulmonary stretch receptors (PSRs) may alleviate the sensation of air hunger.
- The sensation of chest tightness, commonly seen in patients with asthma, probably arises from stimulation of pulmonary receptors as a consequence of bronchospasm.
- The sensation of chest burning associated with acute bronchitis is likely caused by the stimulation of RARs and C fibers in the lungs.
- Dissociation of efferent neural impulses arising within the controller and reafferent neural messages from receptors throughout the respiratory system increases the intensity of dyspnea.
- A patient with a single cardiopulmonary disorder may have multiple physiological abnormalities, each of which may give rise to a qualitatively distinct respiratory sensation.

REFERENCES

1. Simon PM, Schwartzstein RM, Weiss JW, et al: Distinguishable sensations of breathlessness induced in normal volunteers. *Am Rev Respir Dis* 1989, 140:1021–1027.
2. Simon PM, Schwartzstein RM, Weiss JW, et al: Distinguishable types of dyspnea in patients with shortness of breath. *Am Rev Respir Dis* 1990, 142:1009–1014.
3. Chronos N, Adams L, Guz A: Effect of hyperoxia and hypoxia on exercise-induced breathlessness in normal subjects. *Clin Sci* 1988, 74:531–537.
4. Banzett RB, Lansing RW, Reid MB, et al: "Air hunger" arising from increased PCO_2 in mechanically ventilated quadriplegics. *Respir Physiol* 1989, 76:53–68.
5. Banzett RB, Lansing RW, Brown R, et al: "Air hunger" from increased PCO_2 persists after complete neuromuscular block in humans. *Respir Physiol* 1990, 81:1–18.
6. Schwartzstein RM, Simon PM, Weiss JW, et al: Breathlessness induced by dissociation between ventilation and chemical drive. *Am Rev Respir Dis* 1989, 1231–1237.
7. Mahler DA, Harver A, Lentine T, et al: Descriptors of breathlessness in cardiorespiratory diseases. *Am J Respir Crit Care Med* 1996, 154:1357–1363.
8. Moy ML, Weiss JW, Sparrow D, et al: Quality of dyspnea in bronchoconstriction differs from external loads. *Am J Respir Crit Care Med* 2000, 162:451–455.
9. Taguchi O, Kikuchi Y, Hida W, et al: Effects of bronchoconstriction and external resistive loading on the sensation of dyspnea. *J Appl Physiol* 1991, 71:2183–2190.
10. Binks AP, Moosavi SH, Banzett RB, et al: "Tightness" sensation of asthma does not arise from the work of breathing. *Am J Respir Crit Care Med* 2002, 165:78–82.
11. Campbell EJM, Howell JBL: The sensation of breathlessness. *Br Med Bull* 1963, 19:36–40.
12. Lougheed MD, Lam M, Forkert L, et al: Breathlessness during acute bronchoconstriction in asthma. *Am Rev Respir Dis* 1993, 148:1452–1459.
13. Schwartzstein RM, Manning HL, Weiss JW, et al: Dyspnea: A sensory experience. *Lung* 1990, 168:185–199.
14. Manning HL, Schwartzstein RM: Pathophysiology of dyspnea. *N Engl J Med* 1995, 333:1547–1553.
15. Manning HL, Schwartzstein RM. Respiratory sensations in asthma: Physiological and clinical implications. *J Asthma* 2001, 38:447–460.
16. Demediuk BH, Manning H, Lilly J, et al: Dissociation between dyspnea and respiratory effort. *Am Rev Respir Dis* 1992, 146:1222–1225.

Answers TO THOUGHT QUESTIONS

8-1. Dyspnea is a symptom. Respiratory sensations—and any symptom, for that matter—can only be felt by and described by the patient. The son is describing physical signs that he observes and that he assumes are correlated with sensations of uncomfortable breathing. Only the patient can tell you if she is having pain, shortness of breath, or other symptoms. If she consistently denies any respiratory discomfort (and that is the term we use when questioning patients because it is very general and does not presume a particular quality of sensation), she does not have dyspnea.

8-2. Acute respiratory infections, such as pneumonia, often increase the drive to breathe (i.e., the efferent impulses from the controller). This increase in the activity of the controller is likely caused by stimulation of irritant receptors and is often independent of abnormalities in P_aO_2 and P_aCO_2. Because the patient is paralyzed above the cervical levels of the spinal cord that supply the diaphragm and is maintained on a mechanical ventilator, he is unable to respond to the increased efferent output. The response of the ventilatory pump, therefore, is not appropriate to the demands of the controller. In other words, there is efferent–reafferent dissociation, which may be leading to dyspnea. One way to try to alleviate this would be to increase the size of the tidal volume. This will stimulate pulmonary stretch receptors and flow receptors (to increase tidal volume without changing respiratory rate or the proportion of the breathing cycle devoted to inspiration, one has to change the flow), the afferent information from which will still reach the brain despite the spinal cord lesion. Recall from Chapter 2 that the information from pulmonary receptors travels to the brain via the vagus nerve and does not go through the cervical spinal cord. Increasing the respiratory frequency may increase stimulation of flow receptors (to increase the rate with the same tidal volume, flow must increase), but will not give you the additional benefit of the change in afferent impulses from the stretch receptors.

8-3. Studies have shown that the overwhelming majority of patients with asthma attacks report that they experience more difficulty breathing in rather than out. Although we cannot be entirely sure of why this is so, we believe that the answer lies in the fact that exhalation, even during acute bronchospasm, is largely passive. Inhalation, on the other hand, requires a tremendous amount of work in the setting of increased airway resistance and what are typically hyperinflated lungs and chest wall (remember the effect of hyperinflation on compliance of the respiratory system; see Chapter 3 for a review). As the lungs hyperinflate, the respiratory system operates on a flatter portion of the pressure–volume curve (reduced compliance), and the inspiratory muscles are shorter and less effective at generating tension at the beginning of the breath. In more severe episodes of asthma, individuals may also complain of "an inability to get a deep breath" as the end-inspiratory volume approaches total lung capacity.

chapter 8

Review Questions

DIRECTIONS: *Each of the numbered items or incomplete statements in this section is followed by answers or by completions of the statement. Select the ONE lettered answer or completion that is BEST in each case.*

1. The most common physiological cause of respiratory discomfort, or dyspnea, is:

    A. low P_aO_2
    B. high P_aCO_2
    C. problem with the ventilatory pump
    D. problem with the gas exchanger

2. A 45-year-old woman is pulled from a fire. She was trapped in a smoke-filled room for 30 minutes before she was rescued. When paramedics evaluate her, she complains of shortness of breath. Her respiratory rate is 28 breaths/min, the chest examination is free of wheezes, and the oxygen saturation is 82%. On further questioning, the patient is most likely to describe the quality of her breathing discomfort as:

    A. increased urge to breathe; sensation of air hunger
    B. chest tightness or constriction
    C. increased effort to breathe; increased work of breathing
    D. heavy breathing; breathing more

3. A patient who had a right lung transplant 2 weeks ago has a coughing spell while eating and aspirates some food into the right lung (Note: the transplanted lung is severed from the vagus nerve.) You listen to his chest and hear wheezes on the right side. When you ask the patient if he has a sensation of chest tightness, he is likely to say:

    A. yes
    B. no

4. A 62-year-old patient who suffered a fracture of his fourth cervical vertebrae is admitted to the hospital with weakness and shortness of breath. On examination, he is breathing with a respiratory rate of 32 breaths/min and has small tidal volumes. An arterial blood gas shows a P_aO_2 of 70 mm Hg and a P_aCO_2 of 41 mm Hg. The patient is most likely to use which of the following terms to describe the quality of his breathing discomfort?

    A. increased urge to breathe; sensation of air hunger
    B. chest tightness or constriction
    C. increased effort to breathe; increased work of breathing
    D. heavy breathing; breathing more

5. A patient with respiratory failure caused by pneumonia and hypoxemia is being sustained on a mechanical ventilator. You are supplying a fixed tidal volume and inspiratory flow; the patient can determine her own respiratory rate. Although her P_aO_2 is now 90 mm Hg on supplemental oxygen through the ventilator, she

looks uncomfortable, and her respiratory rate is 36 breaths/min. Which of the following changes in the ventilator settings are likely to decrease the patient's breathing discomfort?

A. decrease the tidal volume
B. increase the tidal volume
C. decrease the inspiratory flow
D. increase the inspiratory flow
E. A and C
F. B and D

chapter 9

Exercise Physiology: A Tale of Two Pumps

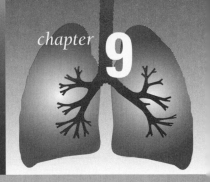

CHAPTER OUTLINE

LEARNING OBJECTIVES

- **To delineate the ways in which the elements of the respiratory system—the controller, ventilatory pump, and gas exchanger—adapt to the metabolic stress of exercise.**
- **To describe the changes that occur in acid–base balance during exercise and the body's adaptations to those changes.**
- **To define the respective roles of the respiratory and cardiovascular systems in supporting the metabolic needs of the body during exercise.**
- **To demonstrate the integrated nature of the physiology of the cardiopulmonary systems in the delivery of oxygen to metabolically active tissue.**
- **To create a framework for understanding exercise physiology in the context of two pumps (respiratory and cardiac) and two gas exchangers (pulmonary capillary and systemic capillary).**
- **To outline the physiological limits of exercise in individuals with normal cardiopulmonary systems.**

You are rounding the corner on the last lap of the mile run. Although you are in the lead, you are being chased by an opponent who is rapidly closing the gap. You are moving large volumes of air into and out of your lungs. You can feel your heart pounding. You are breathing heavily and have a sense of gasping for air. You feel like you just can't go any faster, and you're not sure that you will even be able to make it to the finish line. Why can't you drive your body further? How is the respiratory system responding to the demands placed on it by this intense physical effort? What is limiting you—your respiratory system, or your cardiovascular system?

To this point, we have discussed, in turn, each of the components of the respiratory system. In Chapter 8, we examined how the function of the different elements contributes to respiratory sensations. Now we will investigate the manner in which the three components— the controller, ventilatory pump, and gas exchanger—work together in an integrated manner to enable the body to meet its greatest normal physiological stress: exercise. What you learn in this chapter is also applicable to pathologic states that stress the body in similar ways. Severe infections, for example, often lead to a range of physiologic responses that we term *sepsis* and that are characterized by changes quite similar to what is observed during exercise.

To fully understand exercise physiology, however, and to answer the questions posed in the first paragraph, we must also explore aspects of the function of the cardiovascular system. Those who have already studied cardiovascular physiology may come to appreciate a common framework between the respiratory and cardiovascular systems. Those who have not yet encountered this material before will be introduced to concepts that will have a striking similarity to those you are mastering in your explorations of the respiratory system. If this is your first taste of cardiovascular physiology, do not feel discouraged if you do not fully grasp all of the material. The framework we outline may be of assistance to you when you come to this material more fully in your studies. In the final analysis, exercise physiology is a tale of two pumps—the ventilatory pump and the cardiac pump—and two capillary beds or gas exchangers that provide for the transfer of oxygen from air to blood to metabolically active tissue and a similar transfer of carbon dioxide in the opposite direction.

The Metabolic Demands

As every "couch potato" knows, exercise requires work, generally perceived as moving the body some distance. To accomplish this work, the body must generate energy in the form of adenosine triphosphate (ATP). And because no good deed goes unpunished, the metabolic processes that provide the energy to sustain exercise result in byproducts that, if we are to avoid harm, call upon compensatory responses from the respiratory system.

AEROBIC METABOLISM

During the vast majority of our lives, we rely on aerobic metabolism to provide energy for our tissues. The body uses a combination of carbohydrates and fat to generate ATP, the primary energy currency. Protein may also be used to generate energy, but the body prefers to conserve amino acids for the growth and repair of organs and the enzymatic processes needed to support metabolism.

The amount of carbon dioxide produced as a byproduct of aerobic metabolism varies with the material being used as a fuel source. One is able to compare the different fuel sources in this regard by examining the **respiratory quotient (RQ)**, the ratio of carbon dioxide produced per unit of oxygen consumed in the production of energy.

$$RQ = \frac{\dot{V}CO_2}{\dot{V}O_2}$$

Oxygen consumption is a measure of metabolic activity in the body. You can assess it by measuring the difference between the amount of oxygen that enters the body in a given time interval, usually measured as milliliters per minute, and the amount that is exhaled, also measured as milliliters per minute.

$$\dot{V}O_2 = \text{Amount of oxygen entering the mouth} - \text{Amount of oxygen exhaled}$$

Carbon dioxide production can be expressed in a similar fashion as the difference between the amount of CO_2 exhaled and the amount inhaled.

$$\dot{V}CO_2 = \text{Amount of carbon dioxide exhaled} - \text{Amount of carbon dioxide inhaled}$$

Because there is virtually no carbon dioxide in the gas we inhale from the atmosphere, **carbon dioxide production** is equal to the amount of carbon dioxide we exhale. Substituting 0 for the amount of carbon dioxide inhaled, the equation can be written as follows:

$$\dot{V}CO_2 = \text{Amount of carbon dioxide exhaled}$$

The RQ for carbohydrate is 1.0; for fats, it is 0.7; and for protein, it is 0.8. For someone who eats the average American diet, resting metabolism relies more on the consumption of fat than carbohydrates, and the RQ at rest is approximately 0.8.

THOUGHT QUESTION 9-1: You have a patient with a severe disorder of the ventilatory pump. At rest, she is just able to maintain normal minute ventilation, has a P_aCO_2 of 40 mm Hg (normal), and is mildly short of breath. You now place her on an "anti-Atkins" diet and give her only carbohydrates to eat. What might you expect to happen to her P_aCO_2 and her respiratory symptoms?

During exercise (particularly high-intensity, short-duration exercise), however, the body shifts to greater utilization of carbohydrates, and the RQ changes accordingly with greater production of carbon dioxide.

Normal oxygen consumption at rest is approximately 250 mL/min, and normal carbon dioxide production at rest on a typical mixed diet is roughly 200 mL/min (hence, an RQ of 0.8). During exercise, oxygen consumption may increase to as high as 3000 mL/min (or even higher in athletes), and carbon dioxide production increases even more dramatically as carbohydrate becomes the primary source of energy. The respiratory system, in order to maintain homeostasis, must accommodate these metabolic needs by increasing ventilation.

ANAEROBIC METABOLISM

As the intensity of exercise increases, the body is unable to derive all of its energy needs from aerobic metabolism, and there is a shift to anaerobic processes. Generally, this occurs at a level of exercise that corresponds to 50% to 60% of the individual's maximal oxygen consumption, that is, the oxygen consumption associated with the person's exercise at the point that she says she can go no further (Note: in highly trained competitive athletes, anaerobic threshold [AT] may approximate 80% to 90% of the maximal oxygen consumption.) Tables of predicted values of maximal oxygen consumption based on the individual's size and age can be used for a reference when assessing whether the AT is normal or abnormal. The addition of anaerobic metabolism to supplement energy needs leads to increased demands on the respiratory system.

Aerobic metabolism leads to the production of a byproduct, carbon dioxide, that must be eliminated via the respiratory system. Anaerobic metabolism results in the production of lactic acid. As discussed in Chapter 7, to minimize the effects on the pH of the blood when acid is produced, the body buffers the acid with bicarbonate, a process that ultimately produces carbonic acid, which dissociates to water and carbon dioxide. The anaerobic metabolism of a molecule of glucose, for example, can be summarized as follows:

$$Glucose \rightarrow 2 \ Lactic \ acid + 2 \ ATP$$
$$HLactate + NaHCO_3 \rightarrow NaLactate + H_2CO_3$$
$$H_2CO_3 \rightarrow H_2O + CO_2$$

Thus, anaerobic metabolism leads to the production of additional carbon dioxide via buffering of lactic acid. As with aerobic processes, this process adds an additional stress to the ventilatory pump as it strives to increase ventilation to maintain homeostasis. Anaerobic metabolism, although necessary to support intense exercise, is relatively inefficient compared with aerobic processes. Only two ATP molecules are produced per molecule of glucose metabolized via anaerobic mechanisms. In contrast, the body reaps 36 molecules of ATP for each molecule of glucose metabolized aerobically.

The point during exercise at which the body begins to increasingly rely on anaerobic metabolism to the degree that metabolic acid accumulates is termed the **anaerobic threshold (AT)**. The AT is expressed in terms of the level of oxygen consumption at which lactic acid production can be detected. The point at which AT occurs depends primarily on two factors: the ability of the body to deliver oxygen to the tissues (determined by the function of both the cardiovascular and respiratory systems as well as the hemoglobin content of the blood) and the ability of the tissue to extract and use oxygen to support aerobic processes (determined largely by the density of mitochondria, the quantity of enzymes necessary to support aerobic metabolism in exercising muscles, and the density of capillaries delivering blood and oxygen to the tissue in the muscles). Both of these factors can be altered by your level of *physical conditioning* or *fitness*. As discussed in more detail later in this chapter, oxygen delivery depends on the contractile function of the cardiac pump that, similar to the level of aerobic enzymes in the muscles, responds positively to exercise training and is adversely affected by a sedentary lifestyle.

THOUGHT QUESTION 9-2: It's a beautiful spring day, and you decide to go out for a run. You did not exercise significantly all winter and after only a half mile, you find that you are breathing very hard and have to stop. For the next month, you exercise on a treadmill each day, gradually increasing the speed of the treadmill and the duration of the workout. You now go out again for a run and find that you can go for a mile, but you seem to be breathing less heavily than on your first foray around the track. What has happened to your AT? Is your ventilation less or more on the second run than the first? Why?

The AT can be assessed either by sampling arterial blood repeatedly during exercise to determine the level of oxygen consumption at which lactic acid begins to accumulate or by examining the slope of the graph of ventilation as a function of oxygen consumption during exercise (more about this shortly).

The Respiratory System During Exercise

THE CONTROLLER

In Chapter 6, we began our discussion of the control of ventilation during exercise. Although we can characterize the response fairly accurately at this point, the exact physiological mechanisms responsible for each of the components of the ventilatory response remain the subject of some controversy and speculation.

Phase 1: The Neurological Phase

During the neurological phase of exercise ventilation, breathing increases out of proportion to the metabolic needs of the body. The increase in ventilation, in essence, starts before or accelerates more quickly than the increase in oxygen consumption and carbon dioxide production. Furthermore, the increase in ventilation does not appear to be the consequence of changes in arterial blood gases.

 THOUGHT QUESTION 9-3: What would you predict the acid–base status of a normal person to be during the early stages of exercise?

Studies in animals suggest either a central or peripheral neurological mechanism for this first stage of exercise hyperventilation. Electrical stimulation of areas in the hypothalamus in animals, for example, has been shown to produce respiratory responses that mimic those observed during exercise. Alternatively, experiments in which the limbs of sedated animals were passively moved to simulate exercise demonstrated an increase in ventilation similar to that seen with physical activity. The stimulation of joint or muscle receptors by this movement may trigger neural impulses to the brain that lead to an increase in ventilation.

Some have termed this early stage of exercise ventilation an "anticipatory" response to the metabolic needs that are to follow. It remains unclear whether there is a learned or behavioral component to this response.

Phase 2: The Metabolic Stage

During the second stage of exercise breathing, ventilation increases in concert with the increase in oxygen consumption and carbon dioxide production. The relationship between ventilation and either metabolic parameter is linear. Thus, whether you plot ventilation as a function of carbon dioxide production or oxygen consumption, a straight-line relationship is seen (Fig. 9-1).

 THOUGHT QUESTION 9-4: What relationship exists between the slope of ventilation as a function of carbon dioxide production and the slope of ventilation as a function of oxygen consumption?

The close link between these metabolic parameters and ventilation suggests that the body has a mechanism for monitoring carbon dioxide production or oxygen consumption (or both) and translating that information into a signal to the ventilatory controller. There is no known receptor, however, that serves this purpose. Because P_aCO_2 changes little during this phase of exercise, the chemoreceptors are not likely candidates for the very dramatic increases in ventilation that we observe (easily up to a 10-fold increase over resting ventilation). Some evidence suggests that there may be receptors sensitive to changes in the local metabolic environment at the tissue level located in the skeletal muscles. These so-called **metaboreceptors** could play a role in the increased ventilation during the metabolic phase of exercise. Metaboreceptors are hypothesized to respond to the local accumulation of metabolic byproducts and may send signals to the brain that lead to an increase in ventilation as well as the sensation of shortness of breath with which we associate exercise.

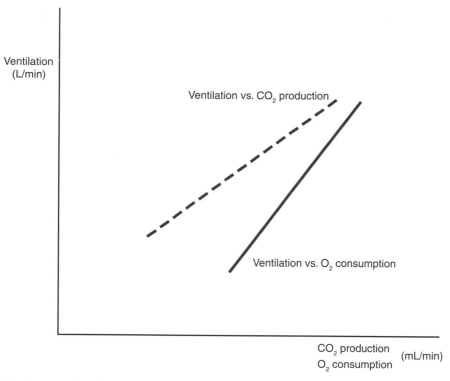

FIGURE 9-1 Ventilation during the metabolic stage of exercise breathing. Ventilation is plotted as a function of oxygen consumption and, secondly, as a function of carbon dioxide production (as exercise intensity increases). In each case, the relationship is linear. This pattern of ventilation during exercise is characteristic of a close link between breathing and metabolic needs, although the exact mechanism by which the body makes these links is not known. During the metabolic phase of exercise, the respiratory quotient increases. This is reflected by the convergence of the lines.

Phase 3: The Compensatory Phase

In our discussion of exercise metabolism earlier in this chapter, we outlined the sequence of events that leads to the production of energy in the muscles. Both aerobic and anaerobic processes lead to the creation of carbon dioxide molecules that must be eliminated via the respiratory system. The exhaled carbon dioxide is what we measure as carbon dioxide production. However, anaerobic metabolism also leads to the production of protons, or acid, that, when buffered by bicarbonate, lead to a further increase in carbon dioxide. Thus, the compensation for the acid that results from anaerobic metabolism involves a further increase in ventilation. At this point, the carbon dioxide exiting the lungs reflects, in part, the consequence of processes that result in the production of energy as well as the compensation necessary to maintain acid–base balance within the body. As protons accumulate and the pH begins to decrease, the peripheral chemoreceptors respond by sending messages to the controller that result in an increase in ventilation.

As one follows the increase in ventilation as a function of oxygen consumption during the metabolic phase of exercise hyperpnea, the slope of the relationship changes as the intensity of exercise increases. Ventilation increases at an even faster rate during the compensatory phase (Fig. 9-2).

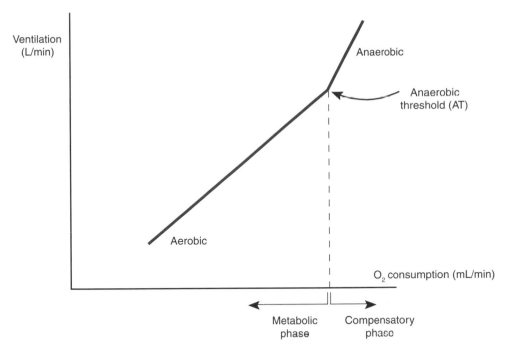

FIGURE 9-2 Ventilation and the anaerobic threshold (AT). Ventilation is plotted as a function of oxygen consumption. The relationship between ventilation and carbon dioxide production (as well as oxygen consumption) is linear through the metabolic phase of exercise hyperpnea. As one enters the compensatory phase of exercise ventilation, however, there is a need to increase ventilation at a greater rate because the body must accommodate the accumulation of metabolic acid. The level of oxygen consumption at which this shift occurs is the AT. Note that we have drawn a somewhat idealized version of this relationship to illustrate our point. In reality, the point at which the slope of ventilation changes is not quite this distinct. **Also see Animated Figure 9-2.**

THOUGHT QUESTION 9-5: What do you predict the P_aCO_2 is during the compensatory phase of exercise ventilation? How would you characterize the acid–base status of the body during this phase?

The change in slope of the relationship between ventilation and oxygen consumption reflects the compensation of the respiratory system for the accumulation of metabolic acid. The level of oxygen consumption at which the slope changes is the AT. The use of the plot of ventilation as a function of oxygen consumption to measure the AT is termed the "V-dot" method, a phrase that signifies that you are using ventilation $\dot V$ to assess the onset of anaerobic metabolism rather than a direct measurement of pH.

 Use Animated Figure 9-2 to vary the level of exercise and observe the effects on parameters, including pH and P_aCO_2. Pay special attention to the changes as you exceed AT and enter the compensatory phase.

THE VENTILATORY PUMP

The controller is stimulated by a variety of factors during exercise, and the demand for increased ventilation is great. The neural messages descend from the controller to the ventilatory muscles, and the pump must respond.

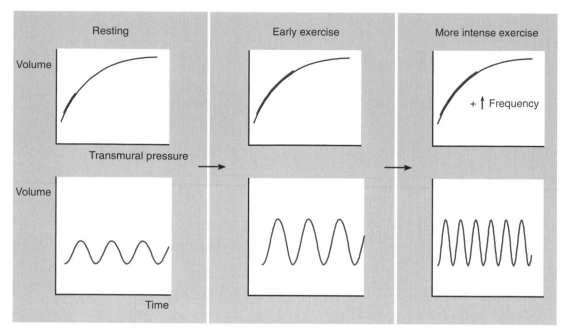

FIGURE 9-3 Exercise ventilation: tidal volume versus frequency. We have plotted the pressure–volume (P-V) relationship for the respiratory system. During resting breathing, the excursion of the respiratory system is along a portion of the curve that is quite compliant, that is, has a relatively high slope. One can increase the volume of the system without generating a very large change in transmural pressure. As exercise begins, tidal volume can increase without moving from this very compliant portion of the curve. However, as tidal volume continues to increase, the end-inspiratory volume now takes you to a flatter portion of the P-V curve; compliance is reduced as the lung and chest wall are stretched farther. The system now responds to additional demands for ventilation by increasing the respiratory rate or frequency of breathing. Assuming that airway resistance is normal, the higher flows that must be generated with an increase in frequency require less work than further enlargement of tidal volume.

The respiratory system may generate the necessary ventilation during exercise by increasing the respiratory rate, increasing the tidal volume, or a combination of the two. The initial increase in ventilation is achieved primarily by enlarging the tidal volume. After doubling the tidal volume, the system relies to a greater degree on changes in respiratory rate to achieve even higher levels of ventilation (Fig. 9-3).

The "choice" between enlarging the tidal volume versus increasing the respiratory rate or frequency to achieve the desired ventilation probably represents an effort by the body to minimize the work of breathing and, hence, the discomfort associated with exercise. Until the tidal volume is doubled, the ventilatory pump operates on a relatively compliant portion of the pressure–volume curve for the respiratory system. (For a review of compliance, see Chapter 3 and Animated Figure 3-5.) Relatively small changes in transmural pressure lead to substantial tidal volumes. As soon as tidal volume has doubled, however, further increases in tidal volume begin to take you to flatter portions of the curve, and compliance decreases. Now it may be more energy efficient to increase ventilation by keeping the tidal volume constant and increasing the frequency of breaths (assuming airway resistance is normal).

  THOUGHT QUESTION 9-6: In a patient with pulmonary fibrosis, a condition in which there is scarring of the lung, causing it to be stiff, how would you expect the respiratory rate during exercise to change compared with the process in an individual without this disorder?

THE GAS EXCHANGER

The absolute amount of dead space in the lungs and the proportion of each breath that is attributed to dead space are both reduced during exercise. You recall from Chapter 5 that a normal individual in the upright position has a small amount of alveolar dead space because of the poor perfusion pressure in the pulmonary capillaries in the apex of the lung (zone 1 of the lung). These regions are ventilated but not perfused because most of the blood flow goes to the bases of the lungs primarily as a consequence of the effect of gravity. With exercise, however, the amount of blood pumped by the heart each minute, the **cardiac output**, increases, and perfusion is more evenly distributed throughout the lung. Previously nonperfused alveoli, therefore, now receive blood, and dead space is reduced.

The proportion of each breath that is composed of dead space ventilation—the ratio of dead space to tidal volume, or V_D/V_T—is also reduced. First, the absolute amount of dead space, as we have just discussed, is smaller during exercise. Second, the tidal volume increases through the early stages of exercise and typically doubles during moderate to very intense activity. For any given amount of dead space, therefore, the V_D/V_T ratio is reduced.

The ability to get oxygen into the blood is also affected by exercise. Capillary blood volume increases during exercise because of the increase in cardiac output. This is favorable for gas exchange because more red blood cells (RBCs) are in contact with the alveolus at any moment in time. However, the increase in cardiac output also results in a shorter period of time that any given RBC is in the alveolar capillary. Recall that at rest, the hemoglobin in an RBC is fully saturated after approximately one third of the transit time through the alveolar capillary. Although this provides substantial reserve capacity at rest, cardiac output may increase fivefold during exercise, thereby leading to a diffusion limitation for gas exchange, i.e., the ability of oxygen to cross from the alveolus to the RBC is limited by the rate of diffusion of the gas (see Animated Figure 5-10). Patients with lung diseases that cause a mild abnormality in the alveolar–capillary interface and prolong diffusion may have normal oxygen saturation at rest, but they desaturate with exercise because of this effect.

The alveolar–to-arterial oxygen gradient, or A-aDO$_2$, tends to widen during heavy exercise. The alveolar PO$_2$ tends to increase with the hyperventilation seen in the compensatory phase of the increased ventilation of exercise, and the arterial PO$_2$ tends to remain constant, thereby resulting in a widened A-aDO$_2$. The lack of increase in the P$_a$O$_2$ is attributable to the reduced level of PO$_2$ in the venous blood returning to the heart during exercise as the metabolically active tissue extracts larger amounts of oxygen from the blood (more on this shortly).

The Cardiovascular System During Exercise

We will now explore some of the basic elements of the cardiovascular system and illustrate how, along with the respiratory system, the cardiovascular system adjusts to the metabolic demands of the body during exercise. Clearly, for those who have not yet embarked on the study of the cardiovascular system, this material will be new and challenging. Nevertheless, for those for whom this is a review, we hope to show you that the structure we have used to understand respiratory physiology (controller, pump, and gas exchanger) can be used as well for the cardiovascular system and that this approach may enhance your understanding of the heart and circulatory system. Of course, we cannot fully discuss all of the physiology of the cardiovascular system within this chapter, but we do provide some basic information that will enable you to see the similarities between the two systems and to understand the physiology of exercise.

BASIC PHYSIOLOGY OF THE CARDIOVASCULAR SYSTEM

The "controller" is the electrical system within the heart and the various neurological and hormonal inputs that modify the rate and force with which the heart contracts. Although all of the cells within the heart have the potential to depolarize and generate an electrical signal, specialized clusters of cells, called the sinoatrial node (SA node) and the atrioventricular node (AV node), act as pacemakers for the heart. Under normal conditions, the SA node initiates an electrical impulse that is disseminated throughout the atria, leading to contraction, and then propagates to the ventricles via the AV node and the specialized conducting tissue known as the bundle of His. This electrical connection allows the contraction of the atria and ventricles to be coordinated to maximize the volume of blood pumped from the heart each minute (i.e., the cardiac output).

Unlike the respiratory system, the cardiovascular system is not subject to voluntary control. One cannot suddenly stop one's heart, even for a few seconds, nor can one command the heart to double its rate of contraction. The pacemaker cells, similar to the inspiratory neurons in the medulla, have an intrinsic firing frequency. This frequency is modified by input from the autonomic nervous system. The two components of the system—the sympathetic and parasympathetic nervous systems—balance each other. Activation of the sympathetic system increases the heart rate (HR), and stimulation of the parasympathetic nervous system tends to slow the frequency of contraction. Output of epinephrine and norepinephrine from the adrenal gland also stimulates β-receptors within the heart to increase the HR. Recall that the sympathetic nervous system also has a role in the modulation of the tone of the smooth muscle surrounding the airways in the lungs. In contrast to the lungs, in which β-2 receptors are stimulated to cause bronchodilation, β-1 receptors are present in cardiac tissue and, when stimulated, are responsible for an increase in HR.

THOUGHT QUESTION 9-7: A young woman comes to see you with a complaint that she suffers from intermittent palpitations during which her heart rate is 130 to 140 bpm. She is healthy except for the fact that she has problems with asthma during the spring and fall allergy seasons and when she gets a respiratory infection. She says that a friend of hers had a similar problem with palpitations and high heart rate and was given a medication called a β-blocker that slowed her heart rate. She wants to know if that would be good for her. What should you tell her?

The autonomic nervous system may be stimulated by emotional factors such as fear or excitement. It may also respond to neurological reflexes designed to maintain blood pressure (BP).

The "pump" component of the cardiovascular system is composed of the heart muscle, the valves within the heart that ensure unidirectional flow, and the blood vessels that form a conduit for flow of blood to the tissues and then back to the right atrium (remember: the cardiovascular *system* is more than just the heart). The heart muscle, similar to the ventilatory muscles, contracts more strongly if it is stretched before receiving a neurological stimulus. The degree of stretch of the ventricular muscle before contraction is termed the **preload**. Thus, the greater the filling of the ventricle before contraction (the greater the preload), the greater the force of the contraction, which leads to a larger volume of blood ejected from the heart (**stroke volume [SV]**) and a larger cardiac output. The force of contraction, or *contractility*, of the heart can also be modified by the autonomic nervous system. Activation of the sympathetic nervous system results in an increase in contractility.

The flow of blood though arteries, capillaries, and veins obeys many of the same physiological rules as does the flow of air through the tracheobronchial tree. For example, the flow, or cardiac output, is determined by the difference in pressure across the circuit (from the aorta to the right atrium) and the resistance of the systemic vasculature (**systemic vascular resistance [SVR]**). Chapter 4 describes the modification of Ohm's law for the flow of fluids through tubes and introduced the following equation:

$$\Delta Pressure/Resistance = Flow$$

When applied to the cardiovascular system, the systemic vascular resistance and cardiac output are substituted into the equation as follows:

$$\Delta Pressure/SVR = Cardiac\ output$$

This relationship is analogous to flow in the lungs, in which the driving pressure is the difference in pressure between the alveolus and the airway opening, and resistance is the airway resistance. The greatest point of resistance in the systemic vasculature is in the small muscular arteries, a location comparable to the point of highest resistance in the airways. The diameter of the muscular arteries and their corresponding resistance is determined by the activity of the autonomic nervous system as well as local factors. In contrast to the lungs, however, where hypoxia leads to vasoconstriction, low oxygen tension in the tissues causes the local systemic vasculature to dilate. This response has the advantage of bringing more blood and, therefore, more oxygen to the hypoxic tissue, thereby preserving aerobic metabolism.

The velocity of the blood (remember from the discussion in Chapter 4 that velocity and flow are not the same; see Animated Figure 4-3) is greatest in the large arteries and lowest in the millions of tiny capillaries arranged in parallel. Similarly, turbulent flow is found in the large arteries, and laminar flow is found in the capillaries. Again, the principles you learned about flow (and Reynolds number) and velocity for the respiratory pump are relevant for the cardiovascular pump.

Finally, the "gas exchanger" for the cardiovascular system is the interface between the capillaries in the tissues and the cells surrounding the vessels. Diffusion of oxygen and carbon dioxide occur in the opposite direction to that seen in the lungs. The principles you learned in Chapter 5 about the binding of oxygen and carbon dioxide to hemoglobin and the factors that affect the binding and release to tissues are all relevant and important here. For example, the accumulation of acid in the tissues from anaerobic metabolism leads to a shift to the right of the oxygen–hemoglobin dissociation curve (see Animated Figure 5-8A). This process reduces the binding of oxygen to hemoglobin and, for any given P_aO_2, permits more oxygen to be available to the tissues.

With this basic information in hand, we will now explore the response of the cardiovascular system to exercise and outline how it works in tandem with the respiratory system to maximize oxygen delivery to the metabolically active muscles.

OXYGEN DELIVERY TO THE TISSUES

Having transported oxygen to the blood, the respiratory system now turns over to the cardiovascular system the responsibility for the next step in the delivery of oxygen to the exercising muscles. **Systemic oxygen delivery** is the total volume of oxygen transported to the tissues each minute. Simply stated, oxygen delivery is the product of the cardiac output (the volume of blood pumped each minute) and the arterial oxygen content (the amount of oxygen per unit volume of blood). This principle is summarized in the following equations.

$$Systemic\ O_2\ delivery = Cardiac\ output \times Arterial\ oxygen\ content$$

If we represent cardiac output with the symbol Qt (total perfusion or flow) and arterial oxygen content as CaO_2, the equation can be rewritten.

$$O_2 \text{ delivery} = Qt \times CaO_2$$

As the blood circulates through capillaries in the tissue, some of the oxygen leaves the blood and enters the surrounding cells. The remainder of the oxygen stays in the blood and returns to the heart and, ultimately, the lungs. The oxygen not used by the tissue returns in venous blood and can be quantified in a manner similar to oxygen delivery. Flow, or cardiac output, in the arterial system must be the same as in the venous system (because in a closed system, if this were not true, there would be a backup). We refer to the venous blood that returns from all of the venous beds, those draining tissue that is very active metabolically as well as tissue that is less active, as **mixed venous blood**. The amount of oxygen returned to the heart can be represented in the following equation:

$$\text{Oxygen returned} = \text{Cardiac output} \times \text{Mixed venous blood oxygen content}$$

The oxygen content of mixed venous blood is represented as CvO_2. The equation can be rewritten as follows:

$$O_2 \text{ returned} = Qt \times CvO_2$$

The difference between the oxygen delivered to the tissue and the oxygen returned to the heart equals the amount of oxygen actually used or consumed by the tissues. Therefore, oxygen consumption may be expressed as:

$$\text{Oxygen consumption} = \text{Oxygen delivered} - \text{oxygen returned}$$

$$\dot{V}O_2 = Qt\,(CaO_2) - Qt\,(CvO_2)$$

$$\dot{V}O_2 = Qt\,(CaO_2 - CvO_2)$$

This relationship is called the **Fick equation** (Fig. 9-4) and is used routinely in clinical practice to calculate the cardiac output in patients in the intensive or cardiac care units as well as in the cardiac catheterization laboratory. Rearranging the equation:

$$Qt = \frac{\dot{V}O_2}{CaO_2 - CvO_2}$$

 Use Animated Figure 9-4 to explore the relationship between cardiac output, oxygen consumption, oxygen delivery, and oxygen extraction in order to better understand the meaning of the Fick equation.

We have already discussed the method for measuring oxygen consumption ($\dot{V}O_2$ is equal to oxygen inhaled minus oxygen exhaled) as well as the calculation of oxygen content of the blood (see Chapter 5). To obtain a sample of mixed venous blood, one needs to have a catheter or tube in the right atrium. The difference between the oxygen content in the arterial blood and venous blood is termed the **A-VO$_2$ difference** (see the vertical axis of Animated Figure 5-8A or 5-8D). In normal individuals, the capacity of the heart to deliver oxygen to the tissues and the ability of the tissues to use oxygen for aerobic metabolism are the limiting factors for exercise. The skeletal muscles have a variable ability to extract oxygen from the blood and use it for aerobic metabolism, depending on the level of "fitness" of the individual. With physical training, biochemical changes occur in the muscle to make it a better "aerobic machine," and the capability of the heart to pump blood with each contraction also improves. Regular exercise also increases the density of capillaries in the muscles, which improves the diffusion of oxygen from the blood to the muscle cells.

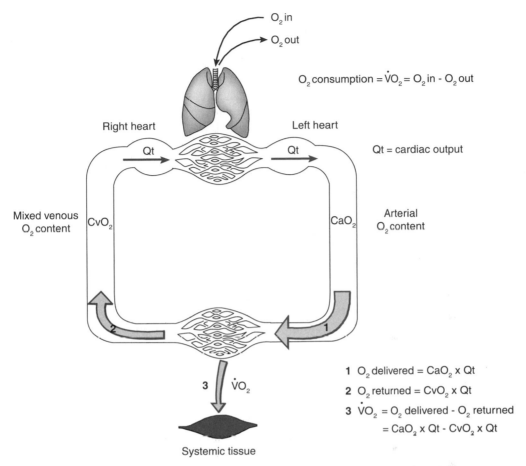

FIGURE 9-4 The Fick equation. Systemic oxygen delivery is the product of cardiac output and arterial oxygen content. The rate at which oxygen is returned to the heart is the product of the oxygen content of mixed venous blood (venous blood found in the large veins, that is, the inferior vena cava or right atrium, representing blood coming back from tissue beds with a range of metabolic activity) and cardiac output. The difference between the oxygen delivered to the tissues and the oxygen returned to the heart is the oxygen consumption. This equals the difference between the amount of oxygen inhaled versus the amount of oxygen exhaled. **Also see Animated Figure 9-4.**

THOUGHT QUESTION 9-8: In a patient with an inability to increase cardiac output, how would the body compensate to support the increased aerobic metabolism associated with exercise?

INCREASING CARDIAC OUTPUT: THE CARDIOVASCULAR CONTROLLER AND PUMP DURING EXERCISE

In a normal person who is resting comfortably, the cardiac output, or volume of blood pumped by the heart, is approximately 5 L/min. During exercise, as the metabolic needs of the body increase, cardiac output may increase fivefold. As we saw with the ventilatory pump, the cardiac pump can increase its output by increasing the volume moved with each contraction or by increasing the rate of contractions (in response to instructions from the controller). Typically, stroke volume doubles during exercise, and the heart rate may nearly triple from resting values (for intense exercise). Again, as discussed in our analysis of the respiratory system during exercise, there are tradeoffs between volume moved per contraction and increases in the rate of contraction.

Stroke volume increases primarily as the result of two physiological mechanisms. First, the more the ventricle fills during *diastole*, the interval between contractions, the greater the strength of the ensuing systole, the contraction phase of the ventricle (recall that the term *preload* describes the initial stretch of the muscle cells of the ventricle prior to contraction). The greater the preload, or ventricular volume at end-diastole, the longer the individual myocardial muscle fibers are. As with the ventilatory muscles, the length–tension relationship of the heart muscle dictates that there is a more forceful contraction with longer muscle fibers. This relationship can be depicted by the **Starling curve**, which displays SV as a function of ventricular end-diastolic volume (Fig. 9-5).

During exercise, the contraction of the muscles of the arms and legs squeezes the veins in the extremities and increases the flow of blood returning to the heart (remember: the flow of blood on the venous side of the cardiovascular circuit must increase in tandem with the increased flow on the arterial side). In addition, stimulation of the sympathetic nervous system leads to the constriction of veins, which further increases the flow of blood back to the heart. The increase in venous flow, termed *venous return*, causes an increase in preload of the right ventricle and, thus, an increase in cardiac output.

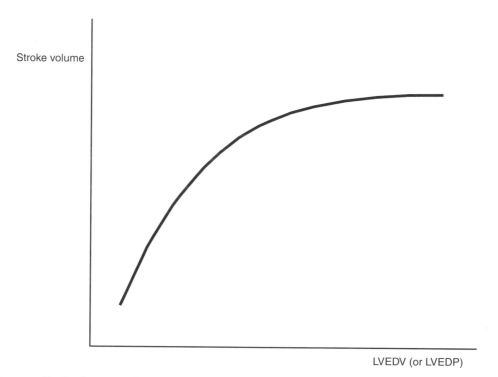

FIGURE 9-5 The Starling curve. The graph depicts the relationship between left ventricular end-diastolic volume (LVEDV) and stroke volume (SV). Note that as the LVEDV increases, so does the SV. This may be viewed as analogous to the length–tension relationship discussed in the context of the contractile function of the ventilatory muscles, but it is now represented in three dimensions because we must conceptualize the contraction of the individual fibers in the context of an organ containing blood. Notice that the curve begins to plateau at higher ventricular volumes, which implies that there are limits to the benefit of stretching the myocardial muscle fibers before the onset of contraction. For a given ventricular compliance, one could also depict this relationship with left ventricular end-diastolic pressure (LVEDP) rather than volume (pressure is easier to measure than volume clinically) because pressure is volume divided by compliance.

As you examine the Starling curve, note that the SV increases as the end-diastolic volume increases only up to a point. Further increases in end-diastolic volume do not yield any greater increases in cardiac output, but they do raise the risk of causing fluid to build up in the lung as the pressure in the pulmonary capillaries increases. Remember that pressure and volume are linked by compliance (in this case, the compliance of the ventricle during diastole).

$$C = \frac{\Delta V}{\Delta P}$$

Increases in end-diastolic volume also cause increases in end-diastolic pressure in the left ventricle. This pressure is transmitted back through the left atrium to the pulmonary veins and, ultimately, the pulmonary capillaries. At a left ventricular end-diastolic pressure (LVEDP) of 18 to 20 mm Hg, one usually begins to see fluid leak into the lungs (the exact pressure at which this occurs is affected by the balance between hydrostatic pressure, the pressure of the fluid in the vessel, and the oncotic pressure, a pressure exerted by the protein elements in the blood; you will learn more about this balance when you study renal physiology). The leakage of fluid into the lungs worsens ventilation/perfusion mismatch and leads to hypoxemia. In healthy individuals, the LVEDP does not exceed 18 mm Hg during exercise, and there is no leakage of fluid into the lungs. In individuals with a stiff or noncompliant ventricle or a ventricle whose ability to pump is severely limited, increases in ventricular pressure are common during exercise, and leakage of fluid into the lungs may occur.

As just described, SV increases during exercise because venous return increases, leading to a greater preload. SV also increases during exercise as a consequence of the effects of the sympathetic nervous system and the chemicals (i.e., epinephrine and norepinephrine) released by the adrenal gland on the contractile function of the heart muscle. For any given preload of the ventricle, the SV will be greater in the presence of increased activity of the sympathetic nervous system. This is visualized graphically by an "upward" shift of the Starling curve (Fig. 9-6).

During exercise, the sympathetic nervous system is "activated," and the SV increases accordingly. This increase in contractility comes with a price, however: sympathetic stimulation of the myocardium increases the oxygen demands of the heart muscle. Under normal circumstances, the heart muscle extracts a very high percentage of the oxygen from the blood traveling through the coronary arteries. If flow in the coronary arteries is reduced, the heart muscle is unable to compensate by extracting more oxygen from the blood that it sees because the percent extraction is already near maximal to start. In individuals with coronary artery disease and limits on the amount of blood and oxygen that can be delivered to the myocardium, exercise can lead to chest pain, heart attack (termed myocardial infarction), and irregular heart rhythms.

Although increasing SV is a very efficient way to increase cardiac output, there are limits, as just discussed, on the ability of the heart to use this strategy to meet the increased metabolic demands of the exercising muscles. Therefore, as we saw with the respiratory system, the body relies on a combination of increased volume per contraction and increased frequency of contractions. As HR increases, oxygen demand of the myocardium also increases, and the same risks to the function and viability of the heart as we saw with increased contractility must be considered. The body ultimately balances these issues by primarily relying on increased SV during mild to moderate exercise and then exhibits greater reliance on HR as the intensity of the activity increases. A well-trained athlete, however, may continue to meet her metabolic needs by increases in SV long after the "couch potato" has shifted to an elevated HR. Physical training strengthens the heart in a way that increases its maximal contractile capability.

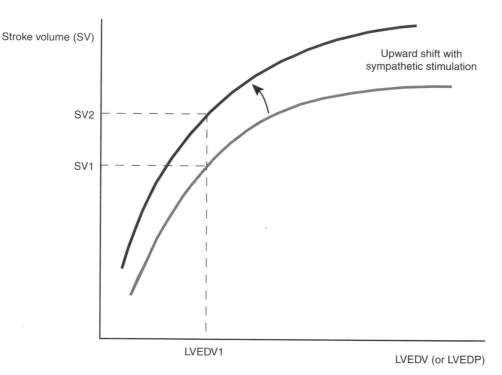

FIGURE 9-6 The effect of the sympathetic nervous system on the Starling curve. Stimulation of the heart by the sympathetic nervous system increases the contractile function of the myocardial cells. In other words, for any given end-diastolic volume, the stroke volume is increased. This is represented by an "upward" shift of the Starling curve. LVEDV = left ventricular end-diastolic volume; LVEDVP = left ventricular end-diastolic pressure.

THE CARDIOVASCULAR GAS EXCHANGER: REDISTRIBUTION OF BLOOD FLOW

The systemic circulation is regulated by the autonomic nervous system and local tissue factors in order to maximize blood flow to the areas that need oxygen and nutrients the most at any particular point. At rest, the muscles are metabolically quiet and receive only 15% of 20% of cardiac output; most blood is sent to the internal organs. During exercise, however, flow is redistributed to the muscles. Stimulation of the sympathetic nervous system causes vasoconstriction, which reduces blood flow to the internal organs. The development of hypoxia locally in the exercising muscle, along with the buildup of the products of metabolism, including lactic acid, cause the arterioles in the muscles to dilate. Arterial resistance is high in blood vessels that serve the internal organs and low in the vessels that supply the muscles. Given the relationship that governs flow:

$$\Delta Pressure = Flow \times Resistance$$

$$\frac{\Delta Pressure}{Resistance} = Flow$$

These differential changes in resistance lead to redistribution of flow (Fig. 9-7). We saw in our study of the respiratory gas exchanger that there are regions of the apex of the lung in an upright person at rest that are poorly perfused or not perfused at all (alveolar dead space) because cardiac output is relatively low at rest. With exercise, cardiac output increases, blood flow to the apex of the lung increases, many pulmonary capillaries are "recruited" to participate in gas exchange, and dead space is reduced. Similarly, as blood flow increases to the muscles, many capillaries that are not perfused at rest are recruited into the gas exchange process. Consequently, the distance between the RBCs in a vessel and the active muscle cells in the limb is reduced, and diffusion is enhanced.

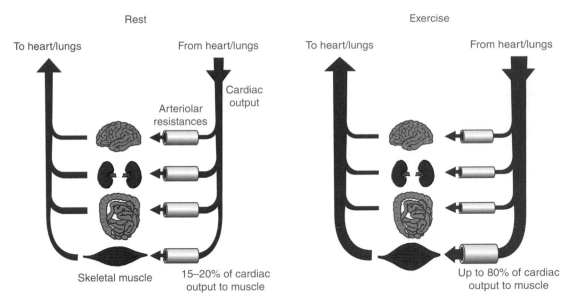

FIGURE 9-7 Exercise and redistribution of blood flow. At rest, the majority of the cardiac output is sent to the vital internal organs. During exercise, the activity of the sympathetic nervous system and local tissue factors in the metabolically active muscles cause blood flow to be redistributed. Vascular resistance increases in the vessels serving the internal organs and decreases in the muscles.

Exercise Physiology: The Concept of Limits

We have been discussing the interaction of two organ systems during exercise as they strive to deliver oxygen to metabolically active tissue, remove carbon dioxide, and maintain an acceptable acid–base balance. These organ systems do an extraordinary job of allowing humans to perform feats of strength and endurance. However, there are limits to what the human body can do. What physiologically prevents us from exercising to even greater levels, the respiratory system or the cardiovascular system?

LIMITS OF THE RESPIRATORY SYSTEM

The ability to deliver oxygen to and remove carbon dioxide from the blood depends on the successful function of all three components of the respiratory system. Adequate ventilation is needed (requiring an adequate controller and pump), and there must not be a significant problem with the gas exchanger in the lungs. Based on exercise studies, we approximate the maximal ventilation that can be sustained in healthy individuals during exercise by multiplying the forced expired volume in 1 second (the FEV_1) by 40. The typical young person has an FEV_1 of roughly 4L. The maximal sustainable ventilation is, therefore, approximately 160 L/min. The highest ventilation reached during exercise rarely exceeds 80 L/min. Ample reserve capacity is available with respect to the functioning of the controller and the ventilatory pump; this is not the limit to exercise.

What about the respiratory gas exchanger? As noted earlier in this chapter, P_aO_2 and P_aCO_2 are normal through most of exercise. Carbon dioxide typically decreases with intense exercise as the respiratory system attempts to compensate for the development of metabolic acidosis. As cardiac output increases and the blood races more quickly through the alveolar capillaries, diffusion of oxygen into the RBCs may not be completed by the time the RBC exits the alveolus. In healthy persons, however, this process rarely causes a significant decrease in P_aO_2 during exercise. Again, ample reserve capacity is available; the gas exchanger is not a limit to exercise.

TABLE 9-1 Indicators* of Physiological Limits During Exercise	
RESPIRATORY SYSTEM LIMITS	**CARDIOVASCULAR SYSTEM LIMITS**
VENTILATORY PUMP	**CARDIOVASCULAR PUMP**
Achieved maximal ventilation ($40 \times FEV_1$)	Maximal HR achieved
Dynamic hyperinflation	Decrease in BP
GAS EXCHANGER	**PUMP OR GAS EXCHANGER**
Development of hypoxemia	Early development of metabolic acidosis (low AT, below
Increase in V_D/V_T	40% predicted maximal oxygen consumption)

*These indicators reflect measurements that are routinely obtained during cardiopulmonary exercise testing.

AT = anaerobic threshold; BP = blood pressure; FEV_1 = forced expired volume in 1 second; HR = heart rate; V_D/V_T = ratio of dead space to tidal volume.

LIMITS OF THE CARDIOVASCULAR SYSTEM

The first sign that there is a problem supporting the needs of the body during exercise is the development of metabolic acidosis. At the AT, the muscle is unable to meet its metabolic needs by aerobic metabolism. What does this tell us physiologically? AT means that there is insufficient delivery of oxygen, or a limited ability of the muscle to use oxygen for metabolism. If the P_aO_2 is normal and the oxygen saturation of the blood is normal, as it should be based on the reserve capacity of the respiratory system, then a limitation on oxygen delivery must be attributable to the cardiovascular system. Once SV has been optimized, cardiac output increases, largely as a result of increasing HR. At the point that the average individual cannot exercise any further, she has reached a limit in cardiac output, typically evidenced by the fact that the HR is greater than 85% of the maximal predicted level (a rough estimate of the maximal predicted HR for a healthy individual can be made by subtracting the person's age from 220). Healthy individuals are limited in their ability to exercise because of the limits of the cardiovascular system in delivering oxygen and the ability of the muscles to use oxygen. Well-trained athletes have more cardiovascular reserve than sedentary individuals, but both groups are limited ultimately by the cardiovascular system (Table 9-1).

PUTTING IT TOGETHER

A 65-year-old woman is referred to you for the evaluation of shortness of breath. She has smoked one pack of cigarettes per day for 40 years and was diagnosed with emphysema 3 years ago. She also suffered a heart attack one year ago. During the past 3 years, she has had gradually worsening breathing discomfort with activity. She has already seen two doctors, one of whom told her that her symptoms were caused by her lung problems; the other attributed them to her heart disease. You ask the patient to describe the quality of her breathing discomfort, and she tells you that she has the feeling that "she cannot get a deep breath" and that she has a sensation of an increased "work to breathe." Based on your understanding of the physiology of respiratory sensations (see Chapter 8 for review), you suspect that she may be developing hyperinflation during exercise because of her emphysema. What might you do to confirm your hypothesis?

To determine the cause of the patient's functional limitation, you perform a cardiopulmonary exercise test, a procedure that assesses multiple parameters relevant to the physiological function of the respiratory and cardiovascular systems during exercise. The patient exercises on a treadmill.

She breathes on a mouthpiece that is connected to a circuit, allowing the analysis of inspired and expired gases. Her HR, BP, electrocardiogram, and oxygen saturation are continuously monitored during the exercise. Pulmonary function tests are performed before and after exercise. Each minute during the test, the patient is asked to take a deep breath in until she cannot breathe in any more. This maneuver allows us to assess changes in the end-expiratory lung volume. If hyperinflation is occurring during the exercise, the inspiratory capacity (i.e., the volume of air that can be inhaled from the end of expiration up to total lung capacity) will diminish. Approximately 10 minutes into the test, the patient indicates that she has to stop. The treadmill is slowed, and the mouthpiece is removed. The patient states that she had to stop because of the same shortness of breath as when she walks too far or rushes to catch a bus. She had no chest pain.

At the point that she stopped exercising, she had reached an oxygen consumption level that was 70% of the predicted maximal level for her age. Her HR was only 65% of her predicted maximum, that is, there appeared to be ample reserve for the heart to increase cardiac output farther. Her BP increased slightly during exercise, as one expects with an appropriate increase in cardiac output. There were no changes in the electrocardiogram to indicate that there was a problem with blood flow to the heart. The oxygen saturation was 97% at rest and decreased only to 94% at the point that the test was terminated. Her maximal ventilation was 45 L/min, exactly 40 times larger than her FEV_1, and the inspiratory capacity maneuvers performed each minute showed that, at the point the patient indicated that she had to stop, the volume of each inhalation during the final stages of exercise was limited by total lung capacity (TLC); she was breathing up to TLC on each breath. How do you interpret these results? What is the cause of her exercise limitation? Why is she not able to go further?

The results of this test indicate that abnormalities in the ventilatory pump were the cause of the patient's exercise limitation. Because of her emphysema, she had increased airway resistance and diminished ventilatory reserve capacity. At the point that she stopped, she had exhausted that reserve. Furthermore, the increased airway resistance during expiration that is the consequence of emphysema caused the patient to develop hyperinflation during exercise; her lung volume at the end of each exhalation was increasing, a form of hyperinflation. This increase in the end-exhalation volume leaves less room for the person to inhale on subsequent breaths. Eventually, the person reached the point that she literally could not take a deep breath because the volume between the end of exhalation and TLC was so small (i.e., the lung volume was so close to TLC). You resolved the dilemma. The exercise limitation was caused by the patient's lung disease.

Summary Points

- The amount of carbon dioxide produced as a byproduct of aerobic metabolism varies with the material being used as a fuel source.
- The respiratory quotient (RQ) is the ratio of carbon dioxide produced per unit of oxygen consumed by the body. The fuel used by the body is a determinant of the RQ. For carbohydrate, it is 1.0; for fats, it is 0.7; and for protein, it is 0.8.
- Oxygen consumption is a measure of the metabolic activity of the body.
- The anaerobic threshold (AT) is the level of oxygen consumption at which the body begins to rely on anaerobic metabolism to meet the energy needs during exercise to such an extent that lactic acid accumulates in the blood.
- As soon as the AT has been reached, the body must compensate for the associated metabolic acidosis that occurs by increasing ventilation above and beyond what is needed to eliminate carbon dioxide produced from aerobic metabolism.
- The AT is determined by the capacity of the respiratory and cardiovascular systems to deliver oxygen to the muscles and by the ability of the muscles to use oxygen to support aerobic metabolism.

- Ventilation increases during exercise in three phases: the neurological phase, metabolic phase, and compensatory phase.
- Ventilation during exercise increases through a combination of enlarged tidal volume and increased respiratory frequency. Healthy individuals will double their tidal volume during moderate to severe exercise. Further increases in tidal volume are energy inefficient because of the reduced compliance of the respiratory system at higher lung volumes.
- The dead space to tidal volume ratio decreases during exercise in healthy persons because of the increase in tidal volume and because there is more even perfusion of the lung (hence, less alveolar dead space) as cardiac output increases.
- The alveolar–arterial oxygen gradient widens slightly during heavy exercise in normal individuals.
- The heart has an intrinsic pacemaker that establishes the rate at which contraction occurs. The pacemaker is under the control of the autonomic nervous system.
- Cardiac output is the volume of blood pumped by the heart each minute.
- The force of contraction of the heart muscle is determined, in part, by the volume of the ventricle at the end of diastole. The stretch of the cardiac muscle cells corresponding to left ventricular end-diastolic volume is the preload of the ventricle. The amount of blood ejected with each contraction of the ventricle is the stroke volume (SV).
- Contractility, the force with which the ventricle contracts, is affected by the autonomic nervous system.
- Systemic arteries (i.e., vessels serving all of the body except the lungs) dilate in the presence of hypoxia in the tissues. This is in contrast to pulmonary vessels, which constrict when exposed to low levels of PO_2 in the alveolus.
- The accumulation of acid in the muscle causes the oxygen–hemoglobin dissociation curve to shift to the right, facilitating the release of oxygen from hemoglobin and subsequent diffusion of oxygen to the tissue.
- Oxygen delivery, the amount of oxygen transported to the tissues each minute, is the product of cardiac output and the arterial oxygen content of the blood.
- The amount of oxygen used by the tissues to support metabolic processes is reflected in the difference in the oxygen content of arterial and venous blood, the A-VO_2 difference.
- The Fick equation allows you to calculate the cardiac output if you know the oxygen consumption and the A-VO_2 difference.
- The heart increases cardiac output by a combination of increased SV and HR. In healthy persons, SV doubles during moderate to heavy exercise, and HR may increase two- to threefold.
- Stroke volume increases during exercise because of increased filling of the ventricle (caused by increased venous return leading to greater preload) and increased contractility (caused by activation of the sympathetic nervous system).
- The Starling curve illustrates the relationship between end-diastolic volume of the ventricle and the SV.
- During exercise, selective constriction of blood vessels serving the internal organs and dilation of arterioles in muscles results in the redistribution of an increased fraction of cardiac output to the muscles.
- During exercise, maximal sustainable ventilation is approximately 40 times the FEV_1.
- Healthy individuals are limited in the amount of exercise they can perform by the cardiovascular system (the ability of the heart to deliver oxygen) and by the ability of the muscles to use oxygen for aerobic metabolism. A person's level of fitness reflects both of these factors.

Answers TO THOUGHT QUESTIONS

9-1. By switching from a diet composed of a mix of food groups to one that contains only carbohydrates, one shifts the patient to a higher RQ (the production of energy from carbohydrates is associated with an RQ of 1.0). Thus, for the same energy requirements, carbon dioxide production increases. In order to maintain a normal P_aCO_2 and pH, the respiratory system must respond by increasing ventilation to eliminate more carbon dioxide. If the patient has a severe disorder of the ventilatory pump, she may not be able to appropriately increase ventilation. Thus, the carbon dioxide level will increase, resulting in the development of acute respiratory acidosis and increased shortness of breath. There is at least one case report of such a patient developing acute respiratory failure because of the initiation of a high-carbohydrate diet.

9-2. Exercise training leads to an increase in the amount of blood the heart can pump each minute (thereby increasing the delivery of oxygen to the muscles) and in the level of aerobic enzymes in the muscle (thereby increasing the amount of oxygen that can be used for aerobic metabolism). The density of capillaries in the muscles also tends to increase with physical training, reducing the distance for diffusion of oxygen from the RBCs to the exercising muscle. Thus, exercise training increases the AT. It is possible that you were beyond AT after a half mile on the first run and had an increase in ventilation that had to compensate not only for excess carbon dioxide resulting from energy production but also for the carbon dioxide resulting from the buffering of lactic acid. After your exercise program, you might be able to run 1 mile relying only on aerobic processes, and ventilation could well be less than on the first run.

9-3. Because the ventilation in early exercise increases faster than the metabolic needs of the body, the neurological phase demonstrates the characteristics of hyperventilation. Specifically, ventilation increases faster than carbon dioxide production. The result is that P_aCO_2 will decrease, and acute respiratory alkalosis will result.

9-4. Oxygen consumption and carbon dioxide consumption are linked via the respiratory quotient (RQ).

$$RQ = \frac{\dot{V}CO_2}{\dot{V}O_2}$$

When the RQ is close to 1, the slopes of the respective curves are nearly the same. As RQ deviates from 1, as for example, in an individual relying primarily upon fat for energy, the slopes become more disparate.

9-5. As soon as the AT has been surpassed, the body increases ventilation, not only to eliminate carbon dioxide resulting from the production of energy but also to compensate for the accumulation of metabolic acid. Thus, hyperventilation is evident, and P_aCO_2 is reduced below normal levels. The individual now has primary metabolic acidosis with compensation by the respiratory system.

9-6. The patient with pulmonary fibrosis has an altered pressure–volume curve for the respiratory system. The curve is flatter (i.e., the slope is less), indicating that the system is less compliant. Greater changes in transmural pressure are required to achieve a given tidal volume. Because these patients generally have normal airway resistance, it is less costly, from the standpoint of the work of breathing, to increase frequency rather than tidal volume to achieve the necessary ventilation required by exercise. These patients, therefore, tend to have higher rates of breathing for any given level of exercise than normal individuals.

9-7. A general or nonspecific β-blocker would block both the β-1 receptors in the heart, leading to a slowing of the HR, and the β-2 receptors in the lung, which could cause bronchoconstriction in a person with a history of asthma. Some β-1 specific blockers are available, however, that allow one to target the receptors contained in the heart without altering the muscular tone of the airways. You tell the patient that because of her asthma, she should only take a β-1 specific blocker for the palpitations.

9-8. To answer this question, reexamine the Fick equation.

$$\dot{V}O_2 = Qt\,(CaO_2 - CvO_2)$$

In order to exercise aerobically, oxygen consumption must increase. If cardiac output is fixed because of an underlying heart problem, the only way to increase oxygen consumption is to extract more oxygen from each unit of blood circulating through the capillaries that supply the muscle. Thus, the difference between the arterial and venous oxygen content of the blood, the **A-VO$_2$ difference**, widens. A large A-VO$_2$ difference is sometimes used as a diagnostic clue to the presence of cardiac disease.

chapter 9

Review Questions

DIRECTIONS: *Each of the numbered items or incomplete statements in this section is followed by answers or by completions of the statement. Select the ONE lettered answer or completion that is BEST in each case.*

1. A patient has an abnormality of mitochondria that minimizes his ability to support aerobic metabolism. You would expect the following to characterize his physiological response to exercise compared with a healthy person.

   A. low anaerobic threshold (AT)
   B. higher than expected minute ventilation for most levels of exercise
   C. dysfunction of the respiratory system controller
   D. A and B
   E. all of the above

2. You are supervising a cardiopulmonary exercise test. You notice that the plot of ventilation as a function of oxygen consumption shows a fairly linear relationship. As the patient continues to exercise, the slope of the plot becomes steeper. You conclude that the patient is now within which phase of the respiratory system controller's response to exercise?

   A. the neurological phase
   B. the metabolic phase
   C. the compensatory phase.

3. A patient with interstitial pneumonitis (an inflammatory condition of the lung that leads to thickening of the alveolar–capillary interface in the lungs) complains of shortness of breath with activity. You assess his oxygen saturation while he is walking up a flight of stairs. You predict you will find the following results.

   A. no change from the resting oxygen saturation
   B. a decrease from the resting oxygen saturation
   C. an increase from the resting oxygen saturation
   D. the oxygen saturation monitor will not work

4. A patient with a cardiomyopathy, a disease that affects the heart muscle and reduces the force of contraction of the ventricles, undergoes a cardiopulmonary exercise test. Assuming he has a normal respiratory system, you predict the following about the level of the AT.

   A. It will be normal.
   B. It will occur at a lower level of oxygen consumption than normal.
   C. It will occur at a higher level of oxygen consumption than normal.
   D. You will not be able to determine where the AT is.

5.   The same patient you assessed in question 4 continues on with his exercise despite passing the AT. As the speed of the treadmill approaches 4.5 mph at an incline of 3%, he complains of dizziness but no chest pain, and you notice that his BP, which had been 150/85 mm Hg, is now 90/60 mm Hg. What physiological events most likely account for the decrease in BP?

A.   The action of the ventilatory pump interfered with the heart's ability to contract.

B.   Venous return was inadequate because his legs became tired.

C.   The accumulation of acid in the blood interfered with the activity of the ventilatory muscles.

D.   The exercise caused the blood vessels in the muscles to dilate, and the heart was unable to increase cardiac output sufficiently to maintain the BP.

Answers to Review Questions

CHAPTER 2: FORM AND FUNCTION: THE PHYSIOLOGICAL IMPLICATIONS OF THE ANATOMY OF THE RESPIRATORY SYSTEM

2-1. **C.** With a damaged motor cortex, the student will not be able to initiate voluntary breaths. Because his brainstem is intact, however, his central pattern generator, located in the medulla, will be functioning. Therefore, assuming there are no residual problems with his gas exchanger because of the drowning and assuming an intact diaphragm, he should be able to resume breathing without the assistance of the ventilator in the future. The cortex is needed to initiate breaths voluntarily, but it is not needed for function of the brainstem pacemaker, which can operate whether the patient is awake or asleep. A phrenic nerve pacemaker is needed primarily in patients with a high (at the level of C1–C2) spinal cord injury that severs the connection between the controller and the ventilatory muscles (see Fig. 2-1 for review).

2-2. **D.** Polio is a disease that affects motor neurons and can lead to paralysis. With weakened ventilatory muscles, the body is unable to generate a sufficient negative intrathoracic pressure to produce a flow of air into the lungs. The vacuum created by the iron lung around the body created a negative pleural pressure and, consequently, a negative alveolar pressure. Because the head (and therefore, the mouth) was outside the iron lung, there was a pressure gradient between the mouth and the alveolus for inspiratory flow. After an appropriate time for inhalation, the iron lung cycled back to atmospheric pressure. The elastic recoil of the lungs then led to the exhalation of air.

2-3. **A.** Whereas the conducting airways have cartilaginous support, the airways in the transitional zone and gas exchanging zone do not. The patient's condition, called tracheobronchomalacia, causes weakening of the supporting cartilaginous structure. Thus, during exhalation, when pleural pressure may be greater than pressure in the airways, the airways tend to collapse. The narrowed airways have a higher resistance than when the diameter is protected by normal cartilage. The transitional zone and the gas-exchanging zone do not contain cartilage.

2-4. **B.** The pulmonary circulation is a low-resistance, high-capacity system. At rest, the volume of blood in the pulmonary capillaries is only one third of the capacity of the vessels. Therefore, the single lung can accommodate the entire output of the heart with very little change in pressure because previously unfilled vessels are recruited into the effective pulmonary circulation. If the patient exercises and cardiac output increases further, increases in pulmonary artery pressure may be seen.

2-5. **C.** When alveolar pressure exceeds the pulmonary capillary pressure, the blood flow to that alveolus diminishes and, in some cases, ceases altogether. Thus, the alveolus is being ventilated, but perfusion is reduced or eliminated. The match between ventilation and perfusion is worsened. The pressure in the pulmonary capillaries varies in the lungs: it is greater at the dependent regions relative to nondependent regions. Thus, the nondependent regions are most affected by high alveolar pressures.

CHAPTER 3: STATICS: SNAPSHOTS OF THE VENTILATORY PUMP

3-1. **D.** Muscle weakness would not lead to changes in FRC, which is determined by a balance of forces between the elastic recoil of the lungs inward and the tendency of the chest wall to spring outward. Conditions that reduce lung compliance tend to increase the recoil of the lung, thereby shifting the balance of forces that determines all three lung volumes in a manner that leads to lower volumes.

3-2. **B.** At a depth of 50 feet, the pleural pressure is greater than at the surface because of the pressure of the water on the chest wall. To inflate the lungs at this depth requires a given pressure that, if the student is holding his breath as he ascends, is the same when he reaches the surface. However, because the pleural pressure decreases as he ascends, the transpulmonary pressure increases, leading to an increase in lung volume.

3-3. **C.** Patients with emphysema have decreased elastic recoil of the lung. This shifts the balance of forces such that TLC and FRC are both increased compared to the normal state. Because FRC is increased, the diaphragm is "lower" in the thorax at the end of a passive breath. This is a "shortened" position of the diaphragm relative to the normal state. Because the tension generated in a muscle depends partly on the length of the muscle at the time it is stimulated, the diaphragm will generate a lower tension than normal in this type of patient.

3-4. **A.** The transpulmonary pressure is the pressure across the lung: $P_{TP} = P_{ao} - P_{pl}$. If the person is holding his breath with the glottis open, there is no flow, and pressure at the airway opening is the same as pressure in the alveolus. Thus, $P_{ao} = P_{alv}$. The size of the lung at any point in time depends on the compliance of the lung and the transpulmonary pressure. Because the lung volume at TLC is greater than at FRC, the transpulmonary pressure is greater at TLC than at FRC. When one holds one's breath at TLC, the pleural pressure is approximately -35 cm H_2O. This results in a transpulmonary pressure of 35 cm H_2O. At FRC, the pleural pressure is approximately -3 cm H_2O, and the transpulmonary pressure is 3 cm H_2O.

3-5. **C.** In conditions in which surfactant is not functioning properly, surface tension in the alveolus increases, resulting in an increased predisposition for the alveolus to collapse. A collapsed alveolus cannot participate in gas exchange. To try to minimize the chance that the alveolus will collapse, one institutes a "positive end-expiratory pressure" with the ventilator. Pressure at the airway opening is not allowed to come down to 0 at the end of exhalation. This results in a greater volume an end-exhalation. By the law of LaPlace, the greater the radius of the alveolus, the lower the pressure in the alveolus required to offset the tension in the wall and thus to prevent alveolar collapse. By maintaining a higher end-expiratory pressure and volume, alveolar collapse is minimized.

CHAPTER 4: DYNAMICS: SETTING THE SYSTEM IN MOTION

4-1. **D.** The patient's emphysema produces areas of lung with very large time constants because airway resistance is high and lung compliance is high. Therefore, it takes a long time for the gas in the alveolus to exchange with the atmosphere. One must wait 15 to 20 minutes after removing supplemental oxygen in such patients before obtaining a measurement of arterial oxygen levels to ensure that the gas in the alveolus no longer has an increased concentration of oxygen from the previous therapy.

4-2. **C.** A patient with emphysema and dynamic airway compression requires a longer time to exhale than an individual with normal lungs. When someone becomes anxious, the usual affect on the breathing pattern is an increase in the respiratory rate. Typically, this leads to a decrease in the time allotted for exhalation. As a result, the patient is unable to get all the air out in the shortened time available (remember: if the patient is flow limited, pushing harder on exhalation will not increase the flow), and hyperinflation results. Hyperinflation is associated with a sensation of not being able to get a deep breath as well as a sense of increased effort to breathe because the muscles of inspiration are shortened at higher thoracic volumes.

4-3. **G.** At higher lung volumes, the patient is able to generate sufficient flow such that he or she can get all the air out that is breathed in despite the limited time for exhalation because of two factors. First, the recoil of the lungs is greater at higher volumes; thus, the driving pressure is greater. Second, the small airways are tethered open by the surrounding lung tissue. At higher lung volumes, the airway diameter is greater; therefore, airway resistance is less.

4-4. **C.** Although the patient's elastic recoil may be above normal because of the pulmonary fibrosis, the fact that TLC is only 50% of predicted means that the driving pressure is likely to be less than in a normal person who is able to inhale up to a normal TLC. In addition, because the patient with pulmonary fibrosis starts the maximal exhalation maneuver at a lower lung volume than normal, the diameter of the airways is reduced, and airway resistance increased, compared with a healthy individual who inhales to a usual TLC.

4-5. **B.** Despite the presence of significant expiratory airflow obstruction, people seem to be bothered more by factors that impair their ability to inhale than exhale. This patient is undoubtedly hyperinflated because of the severity of the expiratory obstruction and the rapid respiratory rate. This places the inspiratory muscles in a shortened and relatively inefficient position at the beginning of each breath. Furthermore, the lungs and chest wall are being moved during the respiratory cycle along the stiffer portion of the pressure–volume curve, that is, the respiratory system is less compliant, and it takes a greater change in pressure to inhale a given tidal volume than when the patient is not hyperinflated.

CHAPTER 5: THE GAS EXCHANGER: MATCHING VENTILATION AND PERFUSION

5-1.

PHYSIOLOGICAL CAUSE OF HYPOXEMIA	A-aDO$_2$ (NORMAL OR INCREASED)	SIGNIFICANT RESPONSE TO SUPPLEMENTAL O$_2$ (Y/N)	PRESENT AT REST (Y/N)
Decreased PIO$_2$	Normal	Y	Y
Alveolar hypoventilation	Normal	Y	Y
Ventilation/perfusion mismatch	Increased	Y	Y
Shunt	Increased	N	Y
Diffusion abnormality	Increased	Y	N

5-2. **A.** To answer this question, you first must calculate the P_AO_2 using the alveolar gas equation. Remember to put 0.35 for the FIO_2 and 60 for the P_aCO_2. This gives you a P_AO_2 of 175 mm Hg. When you subtract the P_aO_2 of 80, the A-aDO_2 equals approximately 95 mm Hg.

5-3.   **F.** The patient has reduced alveolar ventilation, as evidenced by the high P_aCO_2. In this case, it is likely caused by her underlying lung disease (destruction of tissue in the remaining lung from her scarring or fibrosis condition) given the fact that her total ventilation is actually elevated at 8.75 L/min on the ventilator when the ABG was done. In addition, the A-aDO_2 indicates that there must be a gas exchanger problem present as well. The problem is present at rest, making a diffusion abnormality an unlikely explanation. The patient has responded well to supplemental oxygen (if you calculate the expected P_aO_2 on room air, given an A-aDO_2 of 95 [although this value would be somewhat less off of supplemental oxygen], it would be less than 10 mm Hg; thus, she has had a good response to supplemental oxygen), indicating that $\dot{V}$/Q mismatch, rather than shunt, is the primary explanation for her hypoxemia.

5-4.   **C.** The patient has problems both with the ventilatory pump (muscle weakness) and with the gas exchanger ($\dot{V}$/Q mismatch and an increased ratio of dead space to tidal volume). For her to eliminate the carbon dioxide she is producing each minute, given the constraints of the ventilatory pump and gas exchanger, it is best that we allow her P_aCO_2 to increase (assuming that her kidneys can compensate for the respiratory acidosis by producing more bicarbonate in the system; see Chapter 7). With a higher P_aCO_2, she will also have a higher P_ACO_2. This means that each liter of gas coming out of the lungs will contain a greater amount of carbon dioxide than when the P_aCO_2 is 60 mm Hg. Thus, she will not have to breathe as many liters per minute, and her alveolar ventilation will be within the capability of the limits of her respiratory system. Remember the relationship:

$$\dot{V}_A \propto \frac{\dot{V}CO_2}{P_aCO_2}$$

With a higher P_aCO_2, the patient can eliminate the same amount of carbon dioxide each minute with lower alveolar ventilation.

5-5.   **B.** About 90% of patients with pulmonary embolism have an abnormal alveolar–arterial oxygen gradient. Although most of these patients are also hypoxemic, many patients with pulmonary embolism increase their ventilation (as a result of the hypoxemia, shortness of breath, or stimulation of pulmonary receptors; see Chapter 6) and consequently reduce the P_aCO_2. This hyperventilation will also increase the P_AO_2 (recall the alveolar gas equation) and the P_aO_2 for any given A-aDO_2. Thus, the patient may have a pulmonary embolism with an increased A-aDO_2 but normal oxygen saturation. You need an ABG to calculate the A-aDO_2 and determine if a gas exchange problem exists. If it does, you must continue to evaluate the patient for a suspected pulmonary embolism. Remember that a patient can be hypoxemic and have a normal A-aDO_2 (e.g., with hypoventilation), and a patient can have a normal P_aO_2 yet have an elevated or abnormal A-aDO_2. It is the A-aDO_2 that indicates if there is a problem with the gas exchanger.

CHAPTER 6: THE CONTROLLER: DIRECTING THE ORCHESTRA

6-1.   **B.** The patient's ventilation is in excess of her metabolic needs, as evidenced by the low P_aCO_2 level. This situation is characteristic of hyperventilation. The patient's respiratory rate is low to normal, so she is not tachypneic (she has increased her ventilation via large tidal volumes). Hypoventilation refers to reduced ventilation relative to metabolic needs and is characterized by a high P_aCO_2, and hyperpnea describes increased ventilation in proportion to an increase in carbon dioxide production. The tingling experienced by the patient is called *paresthesias* and is precipitated by changes in the level of ionized calcium in the blood associated with the hyperventilation and resulting changes in the pH of the blood.

6-2.   **C.** The upper airway has flow receptors that, when stimulated, have an inhibitory effect on the central controller. Thus, the patient's drive to breathe is reduced, ventilation may decrease slightly, and the work of breathing, especially in light of his increased airway resistance, is less. With an oxygen saturation of 95% without oxygen, there is little change in the oxygen content of the blood with an increase in the FIO_2. The P_aCO_2 will likely increase, not decrease, because of a decrease in minute ventilation from the inhibitory effect of the nasal receptors on the controller and from the Haldane effect (the shift in the carbon dioxide–hemoglobin curve with added oxygen; see Chapter 5). A placebo effect is possible as a contributing factor, but real physiological changes occur with a flow of gas in the nose and upper airway. Research studies that examine the effect of nasal oxygen on respiratory symptoms cannot truly use compressed air (e.g., gas in a tank used for SCUBA diving that has the concentration of oxygen and nitrogen that we usually breathe at sea level) as a "control" for this reason.

6-3.   **A.** Acutely, the output of the controller is increased as a result of stimulation of the peripheral chemoreceptors (hypoxemia). However, to some degree, the alkaline pH attenuates the activity of the peripheral receptor. In addition, the hypocapnia of the blood leads to the diffusion of carbon dioxide from the fluid around the brain into the blood, reducing the activity of the central chemoreceptor. Two days later, although the hypoxemia is the same, the body will have compensated for the elevated pH in the blood and brain. Thus, these inputs, which had been countering the effect of hypoxemia on the peripheral chemoreceptor, will be removed, and the overall output of the controller will increase. Although the content of an individual's diet may alter the respiratory quotient and the amount of carbon dioxide produced for each molecule of oxygen consumed, these effects are relatively small and are unlikely to play much of a role in a person who does not have severe respiratory system problems and chronic hypercapnia.

6-4.   **F.** The woman, who has hypoxemia and hypocapnia, will have stimulation of the peripheral chemoreceptor from the hypoxemia, but the reduced P_aCO_2 will cause carbon dioxide to diffuse from the brain to the blood. This will reduce the activity of the central chemoreceptor and blunt the ventilatory response to hypoxemia. In contrast, the patient with emphysema also has acute hypercapnia with his bronchitis. This stimulates the peripheral and central chemoreceptors and results in a synergistic effect on the ventilation. Although genetic differences in the ventilatory responses to hypoxemia and hypercapnia exist, they do not appear to be gender specific. The Haldane effect describes the shift in the carbon dioxide–hemoglobin dissociation curve that occurs in the presence of oxygen and is not relevant to this discussion.

6-5.   **G.** The P_aO_2 is not sufficiently low to stimulate the peripheral chemoreceptors, and the patient is neither hypercapnic nor acidemic. Interstitial edema is believed to result in stimulation of J receptors (C fibers) in the lungs and contribute to the tachypnea seen in patients with fluid in the lungs. In addition, the pain and shortness of breath experienced by the patient may lead to increased ventilation via volitional control mechanisms.

CHAPTER 7: THE CONTROLLER AND ACID–BASE PHYSIOLOGY: AN INTRODUCTION TO A COMPLEX PROCESS

7-1. **A.** The head injury and the low respiratory rate suggest an abnormality of the respiratory controller. It appears that the patient is hypoventilating, which will result in an increase in the P_aCO_2 and a primary respiratory acidosis.

7-2. **B.** Renal failure leads to a metabolic acidosis because of the accumulation of fixed acids and, in some cases, compromise of the kidneys' ability to reabsorb bicarbonate. The acidosis causes the pH to decrease, which prompts the controller to respond by increasing ventilation, thereby bringing the pH back toward normal. A patient with severe emphysema is less likely to compensate adequately because of problems with the ventilatory pump and the gas exchanger. With diminished respiratory compensation, the 62-year-old patient will have a lower pH than the individual with a normal respiratory system.

7-3. **A.** The pH shows acidemia. The P_aCO_2 level is elevated, which is consistent with respiratory acidosis. The change in pH is consistent with an acute respiratory acidosis, as is the small increase in serum bicarbonate level (the consequence of intracellular buffering of carbonic acid).

7-4. **D.** The pH shows acidemia. The P_aCO_2 is low, which indicates that the acidemia cannot be the result of a respiratory problem; rather, the patient is hyperventilating to compensate for a metabolic acidosis. The low bicarbonate level is consistent with metabolic acidosis.

7-5. **B.** The pH is low, which indicates the presence of acidemia. The P_aCO_2 is high, suggesting respiratory acidosis. The pH is down approximately 0.03 units for each 10 mm change in the P_aCO_2, which is consistent with a chronic respiratory process. The change in the serum bicarbonate level is also of the magnitude one would expect from renal compensation for chronic respiratory acidosis.

7-6. **B.** The pH is high, which indicates the presence of alkalemia. The P_aCO_2 is low, suggesting respiratory alkalosis. The pH is up approximately 0.03 units for each 10 mm decrease in P_aCO_2, which is consistent with a chronic respiratory process. The change in the serum bicarbonate level is also of the magnitude one would expect from renal compensation for chronic respiratory alkalosis.

CHAPTER 8: THE PHYSIOLOGY OF RESPIRATORY SENSATIONS

8-1. **C.** Although gas exchanger problems (and the hypoxemia and hypercapnia that often accompany them) are a significant cause of breathing discomfort in patients with cardiopulmonary disease, increases in airway resistance and decreases in lung and chest wall compliance predominate. Many patients have a combination of gas exchange and ventilatory pump abnormalities. More often than not, fixing the gas exchange problem by administering supplemental oxygen does not relieve completely the breathing discomfort experienced by the patient.

8-2. **A.** The low oxygen saturation reflects hypoxemia, which stimulates the controller and leads to a sensation of air hunger. Chest tightness is most commonly associated with bronchoconstriction; the absence of wheezes argues against this. There is no apparent problem with the ventilatory pump that might give rise to a sensation of increased effort to breathe. The quality "heavy breathing or breathing more" is most often associated with exercise and deconditioning.

8-3. **B.** Although bronchospasm, which is a primary cause of chest tightness, might well be present in this case in the setting of wheezes on physical examination, the afferent pathway by which the information is transmitted from the lungs to the brain to produce the sensation has been severed. Lung receptors transmit information to the brain via the vagus nerve. A transplanted lung is denervated, that is, it does not have neural connections when placed into the recipient. If the sensation of chest tightness associated with bronchospasm arises from stimulation of pulmonary receptors (rapidly adapting receptors and C fibers), this individual will not experience the sensation. Some data suggest that after several years, a transplanted lung may become reinervated and the sensations restored.

8-4. **C.** The nerves emanating from the cervical roots C3–C5 are responsible for activation of the diaphragm. This patient is likely to have diaphragmatic weakness leading to a problem with the ventilatory pump. In order to generate an adequate tidal volume, the patient must increase efferent output from the motor cortex to compensate for the neurological problem. This is perceived as an increased sense of effort or work of breathing. Although the P_aO_2 is below normal, the A-aDO$_2$ is near normal, and the level of hypoxemia is not sufficient to stimulate the controller and produce air hunger.

8-5. **F.** The patient shows evidence of stimulation of the controller. She may not be getting the desired inspiratory flow and tidal volume because those parameters are controlled with the ventilator. This process leads to efferent–reafferent dissociation and increased breathing discomfort. An increase in tidal volume and inspiratory flow might provide a better match between the efferent neural messages and the mechanical output of the respiratory system, thereby reducing dyspnea. Decreases in tidal volume and inspiratory flow are likely to exacerbate breathing discomfort by worsening efferent–reafferent dissociation.

CHAPTER 9: EXERCISE PHYSIOLOGY: A TALE OF TWO PUMPS

9-1. **D.** Patients with mitochondrial dysfunction are unable to sustain aerobic metabolism normally during exercise and quickly develop lactic acidosis. Anaerobic threshold, therefore, occurs at a lower level of oxygen consumption than normally is seen. The acute metabolic acidosis stimulates the respiratory controller, and ventilation increases faster than one would expect during mild to moderate exercise. This process represents a normal compensatory mechanism and not dysfunction of the controller.

9-2. **C.** At the onset of exercise, ventilation increases faster than oxygen consumption and may be a bit irregular. During moderate exercise, ventilation increases linearly with oxygen consumption and carbon dioxide production (the slope may be a bit less steep than during the neurological phase). As soon as anaerobic threshold is reached, the slope of the curve steepens because the controller is now stimulated by the metabolic acidosis associated with anaerobic metabolism. The level of oxygen consumption at which the slope changes is the anaerobic threshold. The patient is now in the compensatory phase of the controller's response to exercise. The controller is compensating for the primary metabolic acidosis by creating a secondary respiratory alkalosis.

9-3. **B.** In a healthy person, there is substantial reserve capacity in the ability to diffuse oxygen from the alveolus into the RBCs traveling through the alveolar capillary. Only with very intense exercise, when cardiac output has dramatically increased and the transit time of the RBCs through the alveolar capillary is greatly reduced, does a diffusion limitation arise. However, in disease states in which the alveolar–capillary membrane is altered and the reserve capacity for diffusion is compromised, relatively mild exercise may be sufficient to cause hypoxemia because of a diffusion problem. The person's lung disease does not interfere with our ability to measure oxygen saturation.

9-4.   **B.** The anaerobic threshold occurs when the delivery of oxygen to the muscles is insufficient to sustain aerobic metabolism and anaerobic processes increase to the point that lactic acid begins to accumulate. In the setting of a normal respiratory system, the onset of anaerobic metabolism is largely related to the ability of the heart to deliver oxygen by increasing cardiac output. If cardiac output is compromised, the muscles reach their limit of aerobic metabolism earlier than normal; in essence, they outstrip the ability of the cardiovascular system to deliver oxygen at a lower intensity of exercise than under normal conditions. Thus, anaerobic threshold occurs at a lower level of oxygen consumption than normal.

9-5.   **D.** The development of hypoxia and acidosis in the leg muscles leads to dilation of the arterioles in the muscles, reducing systemic vascular resistance. Normally, a concurrent increase in cardiac output acts to sustain BP.

$$\text{Pressure} = \text{Flow} \times \text{Resistance}$$

$$\text{Pressure} = \text{Cardiac output} \times \text{Systemic vascular resistance}$$

If cardiac output is unable to increase because of cardiomyopathy, the decrease in systemic vascular resistance will cause a decrease in BP. Actions of the ventilatory pump may interfere with cardiac output if large negative pleural pressure is present during inspiration, a phenomenon that occurs primarily when significant airway obstruction is present. In people with severe obstructive lung disease, high levels of ventilation may be associated with an accumulation of lactic acid within the ventilatory muscles. This local acidosis, however, rarely affects cardiac performance.

Glossary of Terms

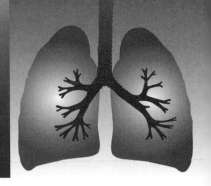

acid: a substance that is capable of donating hydrogen ions (protons or H^+)

acidemia: a condition in which the pH of the blood is below normal

acidosis: a process that leads to the excess production of acid. The ultimate pH of the blood is determined by whether there one or multiple processes are present.

acinus (pulmonary lobule): an anatomical unit of lung distal to a terminal bronchiole; consists of respiratory bronchioles, alveolar ducts, and alveoli. The acinus is the site of gas exchange.

afferent information: neurological sensory messages that arise in the peripheral nervous system and are transmitted to the central nervous system, including information from flow, pressure, stretch, and irritant receptors of the lungs

afferent neural impulse: neural signal generated by the periphery (e.g., receptors in the lungs and chest wall) and sent to the central nervous system (see **afferent information**)

alkalemia: a condition in which the pH of the blood is above normal

alkalosis: a process that leads to excess production of base. The ultimate pH of the blood is determined by whether there one or multiple processes are present.

alveolar gas equation: an equation that describes the relationship between the concentration of inspired oxygen, the P_ACO_2, and the P_AO_2; used to calculate the partial pressure of oxygen within the ideal alveolus

$P_AO_2 = FIO_2 (P_{atm} - P_{H_2O}) - P_ACO_2/R$, where

FIO_2 = fraction of oxygen in inspired gas (0.21 for atmospheric gas)

P_{atm} = barometric pressure (depends on altitude; 760 mm Hg at sea level)

P_{H_2O} = water vapor pressure when gas is fully saturated (47 mm Hg in the air we breathe)

R = respiratory quotient

alveolar-to-arterial oxygen difference (A-aDO$_2$): the difference between P_AO_2 and P_aO_2; determines whether there is a problem with the gas exchanger. The A-aDO$_2$ increases in normal people as they age because of changes in ventilation that relate to loss of elastic recoil in the lungs.

alveoli (singular, alveolus): the thin-walled distal air sacs in the lungs across which gas exchange occurs. Oxygen from the atmosphere is picked up by hemoglobin in the red blood cells traversing alveolar capillaries, and carbon dioxide passes from the blood into the alveoli.

anaerobic threshold: the level of exercise at which the body begins to rely on anaerobic metabolism to meet energy needs such that lactic acid accumulates in the blood; expressed in terms of the level of oxygen consumption at which lactic acid production can be detected

anemia: a condition in which the amount of hemoglobin in the blood is below normal. Anemia reduces the oxygen content of the blood.

anion gap: the difference in concentrations between the commonly measured anions and cations in the blood. The anion gap can be altered by certain acids; thus, calculation of the anion gap serves as an aid to recognition and diagnosis of metabolic acidosis.

arterial blood gas: a test in which a sample of arterial blood is analyzed to determine the P_aO_2, P_aCO_2, and the pH of the blood

atelectasis: phenomenon that involves the collapse of lung units

A-VO$_2$ difference: the difference between the oxygen content in the arterial blood and in the venous blood

base: a substance that is capable of accepting hydrogen ions (protons or H^+)

body plethysmography: a method of measuring functional residual capacity. In this test, a person is seated within an airtight box. The body plethysmograph makes use of Boyle's law to calculate the functional residual capacity based on changes in volume and pressure as the person attempts to inhale and exhale against a closed valve. This method, unlike helium diffusion, is not affected by poorly ventilated regions of the lungs.

bradypnea: abnormally slow breathing; a decrease in respiratory rate below the normal range (usually reserved for rates < 10 breaths/min)

bronchial circulation: delivery of oxygenated blood from the aorta to the lung tissue; serves as a major supply of oxygen and nutrients to the trachea, bronchi, and the visceral pleura (as well as the esophagus)

buffer: a substance that is able to accept or release hydrogen ions, resulting in minimal changes in the free hydrogen ion concentration and, hence, the pH. The primary intracellular buffers are proteins, phosphates, and hemoglobin in red blood cells, and the primary extracellular buffer is bicarbonate.

carbon dioxide content: the amount of carbon dioxide bound to hemoglobin plus the amount dissolved in the blood

carbon dioxide production ($\dot{V}CO_2$): the rate at which carbon dioxide is produced by the body. $\dot{V}CO_2$ can be expressed as the difference between the amount of CO_2 exhaled and the amount inhaled. Because there is virtually no CO_2 in the gas humans inhale from the atmosphere, $\dot{V}CO$ is equal to the amount of CO_2 humans exhale each minute.

cardiac output ($\dot{Q}t$): the volume of blood pumped by the heart each minute

central pattern generator: the area of the brain, including the medulla, which is responsible for automatic control of respiration; it creates the inherent rhythm of breathing

C fibers: unmyelinated fibers in the lungs that carry information from a variety of receptors, including J receptors, and conduct sensory information that arises from inhalation of irritant substances. C fibers may also arise from receptors in the bronchi. The exact role of C fibers in the modulation of respiratory control is unknown.

chemoreceptors: specialized sensory bodies that contain nerve fibers that extend to the medulla in the region where inspiratory neurons are located; the chemoreceptors monitor oxygen and carbon dioxide levels as well as the pH of the blood. Two kinds of chemoreceptors are:

- **central (medullary) chemoreceptors:** the chemoreceptors located in the medulla that are sensitive to changes in carbon dioxide and pH

- **peripheral chemoreceptors:** the chemoreceptors located in the aortic arch and in the carotid bodies at the bifurcation of the common carotid artery; they respond to hypoxemia, hypercapnia, and changes in pH in the blood

compliance: a measure of the stiffness of a closed container such as a balloon; the change in volume that occurs in the object divided by the change in pressure across the wall of the object. Compliance is mathematically expressed as ΔVolume/ΔPressure.

conducting airways: respiratory passages through which air flows but no gas exchange takes place

corollary discharge: the simultaneous message that is sent to the sensory cortex as efferent (outgoing) neural messages are sent from the motor cortex to the ventilatory muscles. These messages from the motor cortex to the sensory cortex are believed to be responsible for the sense of effort, or perception of increased work of breathing, when ventilatory muscle activity is increased.

dead space: regions of the lungs (including the conducting airways) that receive air but do not participate in gas exchange (see **physiological dead space**)

dead space–to–tidal volume ratio (V_D/V_T): a ratio used clinically to assess gas exchange. In an average person, a normal value is 175/500, or about one third. A V_D/V_T ratio greater than or equal to 0.6 is indicative of respiratory failure.

driving pressure: the pressure change from one point to another along a tube through which a fluid flows. Driving pressure, or pressure decrease, is equal to flow multiplied by the resistance of the tube.

dynamic compression: the reversible narrowing of airways that results from changes in transmural pressure during exhalation

dynamics: the study of the respiratory system during conditions of flow of gas into and out of the lungs

dyspnea: breathing discomfort

efferent neural impulse: neural signal generated by the central nervous system (e.g., the respiratory controller) and sent to the periphery (e.g., receptors in the respiratory muscles)

efferent–reafferent dissociation: mismatch of outgoing (efferent) signals to the ventilatory pump and returning (reafferent) signals from receptors in the lungs and chest wall; increases the intensity of sensations such as "air hunger" and the effort or work of breathing

effort independence: condition that exists when increases in pleural pressure do not yield increased expiratory flow (see **flow limitation**)

elastic properties: the properties of a material or object that allow it to return to its original size and shape after being stretched or compressed. Both the lungs and the chest wall have elastic properties.

elastic recoil: in reference to the lungs, the force attributable to the elastic properties (e.g., attributable to stretching of elastin and collagen fibers) that act to return the lungs to their resting or unstressed position. The elastic recoil of the lungs provides a driving force for expiration.

equal pressure point (EPP): the point at which transmural pressure of the airway is 0, that is, the pressure inside and outside are the same. When the EPP is reached in peripheral, unsupported airways, further increases in pleural pressure do not result in increased flow (see **flow limitation**).

expiratory coving: the concave-up appearance of the expiratory flow–volume curve from the point of peak flow to that of residual volume; results from the rapid decrease of expiratory flow at lung volumes below total lung capacity attributable to narrowing of the airways, as in emphysema

expiratory reserve volume: the maximal volume of air that can be expelled from the lungs after a normal expiration; the difference in volume between the functional residual capacity and the residual volume

Fick equation: the relationship between cardiac output (Qt), oxygen consumption ($\dot{V}O_2$), and arterial–venous oxygen content difference (A-VO_2 difference). The oxygen consumption can be expressed as the difference between the oxygen delivered to the tissues and oxygen returned to the heart and is equal to the difference between the oxygen inhaled and the oxygen exhaled each minute. The Fick equation is used routinely in clinical practice for calculation of cardiac output. The equation for cardiac output is:

$$Qt = \frac{\dot{V}O_2}{A\text{-}VO_2 \text{ difference}}$$

fixed acid: a nonvolatile acid, meaning, one that is unable to be excreted by the lungs. Fixed acids include the phosphates and sulfates produced by the metabolism of certain proteins. A functioning renal system is required for the elimination of fixed acids.

flow limitation: a condition characterized by an inability to increase flow at a given lung volume with greater expiratory efforts. When flow limitation is present, equal pressure point conditions exist, and the driving pressure that determines flow is the difference between alveolar pressure and pleural pressure; specifically, driving pressure at flow limitation is the elastic recoil pressure of the alveolus at that lung volume.

forced expiratory volume in 1 second (FEV$_1$): the volume of gas that is exhaled in the first second of the forced vital capacity maneuver. People who have obstructive lung diseases such as emphysema and chronic bronchitis typically have an abnormally low FEV$_1$.

forced vital capacity (FVC): the vital capacity measured with the subject exhaling as rapidly as possible

functional residual capacity (FRC): the volume of the lungs at the end of a normal, relaxed exhalation. At the FRC, the inward or deflating force of the lungs is balanced by the outward or expanding force of the chest wall.

gas exchanger: the alveoli and the pulmonary capillaries that are the site of exchange of oxygen and carbon dioxide. Oxygen from the atmosphere is picked up by hemoglobin in red blood cells traversing the alveolar capillaries, and carbon dioxide passes from the blood into the alveoli.

helium dilution: a technique commonly used to measure functional residual capacity (FRC). In this test, a person breathes into a tube connected to a container with a known volume and an initially known concentration of helium. By monitoring the concentration of helium as it distributes and comes to equilibrium across the combined volume of container plus lungs, the volume of the lungs (initially at FRC when connected to the container) can be calculated. This technique may lead to inaccurate measurements of lung volume when disease states characterized by regions of poorly ventilated lung are present.

hyperpnea: increased ventilation that meets the metabolic needs of the body, as reflected in the production of carbon dioxide. If ventilation is increasing in proportion to increases in carbon dioxide production, P_aCO_2 remains normal.

hyperventilation: increased ventilation greater than that required to meet metabolic needs, as reflected in the production of carbon dioxide, or increased alveolar ventilation relative to metabolic carbon dioxide production. If a person is hyperventilating, it means that his or her P_aCO_2 is less than normal, or less than about 38 mm Hg.

hypoventilation: decreased ventilation to a level that is less than that required to meet metabolic needs, as reflected in the production of carbon dioxide, or decreased alveolar ventilation relative to metabolic carbon dioxide production. If a person is hypoventilating, it means that his or her P_aCO_2 is greater than normal, or greater than about 42 mm Hg.

hypoxemia: a condition in which the PO$_2$ of arterial blood is below normal

hypoxia: a condition in which the level of oxygen in inspired gases or tissues is below normal

hypoxic pulmonary vasoconstriction: narrowing of precapillary pulmonary vessels in response to hypoxia in corresponding poorly ventilated alveoli. This phenomenon aids in directing blood flow preferentially to well-ventilated regions of the lungs.

hysteresis: the separation of the inflation and deflation segments of the pressure–volume curves of the lungs. In other words, the pressure corresponding to a given volume is different for inflation versus deflation. This property is characteristic of air-filled lungs and can be eliminated in experiments in which the lungs are filled with saline (suggesting that hysteresis is caused by the surface forces associated with an air–liquid interface).

inspiratory capacity: the maximal volume of air that can be inspired after a normal expiration. The inspiratory capacity is the sum of the tidal volume and the inspiratory reserve volume.

intracavitary pressure: the pressure inside a closed container or object. In conjunction with the pressure outside the container, this pressure can be used to calculate the transmural pressure, which determines the volume of the object.

J receptors: a group of C fibers that arise deep within the lungs; these juxtacapillary receptors (hence their name) are located near the pulmonary capillaries. J receptors may be activated by increases in pulmonary capillary pressures or accumulation of interstitial fluid, as seen in individuals with congestive heart failure.

laminar flow: a condition in which the movement of fluid molecules in a tube is parallel to the direction of flow. Under conditions of laminar flow, changes in driving pressure are proportional to changes in the flow.

law of Laplace: the relationship between transmural pressure, wall tension, and radius of curvature of a concave surface. This law is mathematically expressed for a sphere as $P = 2T/r$, where P is the pressure inside the sphere, T is the tension in the wall of the sphere, and r is the radius of the sphere.

load compensation: the process in which the controller, the centers in the brain that determine the rate and depth of breathing, makes an adjustment to the presence of increased airway resistance or a sudden change in the compliance of the respiratory system by instructing the muscles to contract more forcefully

mechanoreceptors: specialized sensory bodies that are activated by mechanical distortion of the local environment

medulla: the most caudal region of the brainstem, immediately continuous with the spinal cord. It contains neurons with both inspiratory and expiratory activity and is the primary source of automatic respiratory rhythm (also referred to as the medulla oblongata; see **central pattern generator**).

metaboreceptors: receptors that may play a role in stimulating ventilation during the metabolic phase of exercise. It is hypothesized that these receptors respond to local accumulation of metabolic byproducts in exercising skeletal muscles and send signals to the brain, leading to an increase in ventilation as well as the sensation of shortness of breath with which exercise is associated.

micelles: subsurface aggregates of surfactant molecules; formed when the alveolar radius diminishes and surfactant is extruded from the surface layer

mixed expired gas: a combination of gases from perfused alveoli, from nonperfused alveoli (alveolar dead space), and from anatomic dead space; exhaled gas collected from a person over several minutes. Mixed expired gas is used to measure physiological dead space.

mixed venous blood: venous blood returning to the heart from all tissue beds. To accurately obtain a mixture of all blood returning to the heart, this blood must be drawn from the pulmonary artery.

muscle spindle: a component of skeletal muscle that responds to mechanical stimuli (contraction) and behaves like a slowly adapting receptor in the chest wall; the muscle spindle plays a role in monitoring the response of the ventilatory pump to neural output from the central controller.

nitrogen washout test: a method for measuring functional residual capacity (FRC). The subject breathes a gas containing no nitrogen (typically 100% oxygen), and the volume and decreasing concentration of nitrogen in the exhaled gas are measured. The volume of the lungs (initially at FRC when the test is started) can be calculated in a similar manner to the helium dilution test and has the same limitations as the helium dilution method for assessing functional reserve capacity.

oxygen consumption ($\dot{V}O_2$): the rate at which oxygen is used by the body; a measure of metabolic activity and often given in mL O_2/min. $\dot{V}O_2$ can be expressed as the difference between the amount of O_2 inhaled and the amount of O_2 exhaled.

oxygen content: the sum of the oxygen bound to hemoglobin and dissolved in the blood

oxygen saturation: the percentage of oxygen-binding sites on hemoglobin that are occupied by oxygen; the extent to which the entire oxygen-carrying capacity of hemoglobin is being used

parasympathetic nervous system: along with the sympathetic system, part of the autonomic nervous system; innervates the airways via cholinergic receptors. Stimulation of this system causes bronchoconstriction.

parietal pleura: the thin layer of membranous tissue lining the inside of the chest wall

perfusion: blood flow to an organ or tissue bed. In the lungs, the distribution of perfusion can be influenced by multiple factors, including localized hypoxia and gravity.

physiological dead space: all parts of the lungs (including the conducting airways) that receive air but do not participate in gas exchange; the total dead space. Physiological dead space is the sum of the anatomic dead space and the alveolar dead space, and it is about 175 mL in the average person.

- **anatomic dead space:** the volume of the conducting (non–gas-exchange) airways of the lungs (about 1 mL/lb or 150 mL in the average person)

- **alveolar dead space:** the volume of the alveoli that are not being perfused (about 20 to 50 mL in a healthy person)

pleural space: the potential space between the visceral pleura and the parietal pleura. This space normally only contains a few milliliters of fluid.

pons: a portion of the brainstem superior to the medulla containing neurons that play a role in respiration

pores of Kohn: microscopic passages that connect alveoli within a lobe of the lung, permitting transfer of gas between alveoli. These passages function to minimize the collapse of lung units if a more central airway is obstructed.

preload: the degree of stretch of the ventricular muscle before contraction of the heart

pulmonary circulation: the blood vessels that deliver deoxygenated blood from the right ventricle to the lungs and oxygenated blood from the alveoli to the left atrium

quasistatic: conditions that reflect changes in volume that occur so slowly as to be essentially static. Under these conditions, elastic recoil of the respiratory system is the primary force that must be overcome to achieve changes in volume (i.e., airway resistance is negligible).

rapidly adapting stretch receptors (RARs): myelinated stretch receptors in the lungs that change their firing rate quickly when they reach a new level of stretch; that is, they quickly resume their baseline rate of firing after the change in condition

reafferent: the afferent messages from receptors in the peripheral nervous system (e.g., in the lungs and chest wall) that are generated in response to efferent signals (e.g., from the motor cortex to the ventilatory muscles)

recruit: to make active, as when expiratory muscles are activated (recruited) or when blood vessels in the lung that are normally closed are opened up, making the alveoli they perfuse active participants in gas exchange

residual volume (RV): the volume remaining in the lungs at the end of a maximal expiration

respiratory bronchioles: the smallest bronchioles; these connect the terminal bronchioles to the alveolar ducts. The respiratory bronchioles have alveoli sporadically originating from their walls and are therefore the first site of gas exchange as air travels distally in the lungs.

respiratory controller: the elements of the central nervous system that tell humans how often and how deeply to breathe. These elements include a volitional component in the cerebral cortex and an automatic component in the brainstem. Both of these elements may be affected by a variety of stimuli from throughout the respiratory system.

respiratory failure: the inability of the respiratory system to sustain the metabolic needs of the body

respiratory quotient (RQ): ratio of carbon dioxide produced per unit of oxygen consumed by the body. The fuel used by the body determines this ratio; for carbohydrates, it is 1.0; for fats, it is 0.7; and for protein, it is 0.8. The RQ can be written as:

$$RQ = \frac{\dot{V}CO_2}{\dot{V}O_2}, \text{ where}$$

$\dot{V}CO_2$ = rate of carbon dioxide production

$\dot{V}O_2$ = rate of oxygen consumption

respiratory zone: the portion of the lungs containing alveoli

shunt: a condition in which deoxygenated blood in the venous system is directed to the arterial system without receiving oxygen from the lungs. A pulmonary shunt is an extreme form of ventilation/perfusion mismatch in which some of the alveoli of the lung completely collapse or fill with fluid. With a shunt, alveolar ventilation in the affected regions of the lung is 0. Conditions in which significant shunt is present are characterized by hypoxemia that does not respond to supplemental oxygen.

slowly adapting stretch receptors (SARs): myelinated stretch receptors in the lungs that alter their level of activity in response to a change in local conditions and maintain that new level of activity, or rate of firing, for an extended time after the new conditions have been established (i.e., they are slow to adapt to the new condition and reestablish their baseline firing rate)

spirometer: a device that measures the volume of air inspired or expired. One type, a water spirometer, is a device into which a subject breathes in a manner such that the expired gas is collected under a drum partially suspended under water. As the person inhales and exhales, the water spirometer transforms the motion of the drum into a graphical representation of the change in lung volume as a function of time.

Starling curve: the relationship between the stroke volume and the end-diastolic volume of the ventricle. In general, greater end-diastolic volume leads to increased force of contraction and therefore greater stroke volume and cardiac output.

statics: the study of the respiratory system when there is no air moving into or out of the lungs

stress adaptation: a phenomenon in which elastic recoil forces adjust over seconds after a change in volume of the lung; may play a small role in hysteresis of the lungs

stroke volume: the volume of blood ejected from a ventricle of the heart in a single beat; related to preload (filling or stretch of the ventricle before contraction) via the Starling curve, as well as other factors, including cardiac contractility (see **Starling curve**)

surface tension: the force with which the surface of a liquid contracts per unit length of surface; expressed in dynes/cm^2. Surface tension occurs when a gas–liquid or an air–liquid interface is present and is caused by cohesive forces between molecules in the liquid surface layer.

surfactant: a substance that lowers surface tension within the lungs; produced by type II pneumocytes within the alveoli and becomes a component of the liquid layer lining the alveoli, thereby lowering surface tension and contributing to alveolar stability

sympathetic nervous system: along with the parasympathetic system, part of the autonomic nervous system. In the lungs, stimulation of sympathetic nervous system fibers, which terminate near the airways, causes bronchodilation.

systemic oxygen delivery: the total volume of oxygen transported to the tissues each minute; can be expressed as the product of the cardiac output and the arterial oxygen content

systemic vascular resistance: the resistance of the systemic (i.e., nonpulmonary) blood vessels; small muscular arteries (known as arterioles) represent the site of greatest resistance in the systemic circulation

tachypnea: rapid breathing; an increase in respiratory rate above the normal range (usually reserved for rates ≥20 breaths/min)

terminal bronchioles: subdivisions of the bronchi that represent the last portions of the air conduction pathway before the site of gas exchange

tidal volume (V$_T$): the volume of air inspired or expired in a normal breath. In the average person at rest, a normal tidal volume is approximately 450 to 500 mL.

time constant: a measure of how rapidly gas is moved into and out of the alveoli; provides information about the rate at which gas is replenished within an alveolus. The time constant may be expressed as the product of airway resistance and lung compliance; higher time constants are characteristic of poorly ventilated regions of the lungs.

total lung capacity (TLC): the volume of air in the lungs at the end of a maximal inspiration. The TLC is the sum of the residual volume and vital capacity.

transitional (marginally laminar) flow: generally laminar movement of fluid that involves generation of eddies at angles to the main direction of flow. Transitional flow is typically seen at branch points in tubes.

transitional zone: the zone marking the transition between the conducting airways and alveoli; consists of airways called the respiratory bronchioles

transmural pressure (P$_{TM}$): the pressure across the wall of a structure; mathematically expressed as $P_{TM} = P_{inside} - P_{outside}$, where P_{inside} is the pressure on the inside of an enclosed structure (also known as intracavity pressure), and $P_{outside}$ is the pressure on the outside the structure. Positive transmural pressures tend to expand or increase the volume of a structure, and negative transmural pressures tend to decrease the size or volume of a structure.

turbulent flow: a condition in which the movement of fluid molecules in a tube are in seemingly random directions relative to the main direction of flow. Under conditions of turbulent flow, changes in driving pressure are proportional to the square of the flow.

type II pneumocytes: alveolar cells interspersed among type I pneumocytes. Type II pneumocytes produce, absorb, and recycle surfactant.

vagus nerve: an important nerve that conveys information from sensory receptors in the lungs and other organs to the brain

ventilation ($\dot{V}$): movement of air into and out of the lungs. There is relatively more ventilation than perfusion at the lung apices. Total ventilation ($\dot{V}_E$) is the sum of the alveolar ventilation and the dead space ventilation.

- **alveolar ventilation ($\dot{V}_A$):** the volume of air per minute that enters (or exits) alveoli that are perfused. This air is able to participate in gas exchange.

- **dead space ventilation ($\dot{V}_D$):** the volume of air per minute that enters (or exits) the parts of the lung that do not participate in gas exchange.

ventilation/perfusion ($\dot{V}/Q$) mismatch: an imbalance between alveolar ventilation and pulmonary capillary blood flow. Many disease states, including pneumonia, heart failure, asthma, and emphysema, can increase $\dot{V}/Q$ mismatch and lead to impaired gas exchange.

ventilatory pump: structures that perform the bellows function of the respiratory system; includes the muscles, bones, cartilage, and related soft tissue of the chest wall necessary to move the chest wall, the peripheral nerves that innervate the chest wall muscles, the conducting airways, and the pleura connecting the motion of the chest wall to the lungs

visceral pleura: the thin layer of membranous tissue surrounding the lungs

vital capacity (VC): the maximal amount of air that can be moved into or out of the lungs in a single breath. The VC is the difference between the total lung capacity and the residual volume.

volatile acid: acid that is able to be removed by elimination of CO_2 via the respiratory system; generally refers to carbonic acid. Volatile acids are the products of the metabolism of carbohydrates and fats.

Index

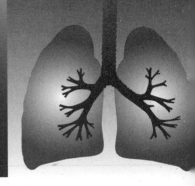

Pages followed by an "f" denote figures; those followed by a "t" denote tables.